RAND NATIONAL DEFENSE RESEARCH INSTITUTE

T0308648

# Defining the Roles, Responsibilities, and Functions for Data Science Within the Defense Intelligence Agency

Bradley M. Knopp, Sina Beaghley, Aaron Frank, Rebeca Orrie, Michael Watson

Prepared for the Defense Intelligence Agency

For more information on this publication, visit www.rand.org/t/RR1582

**Library of Congress Cataloging-in-Publication Data** is available for this publication.
ISBN: 978-0-8330-9658-6

Published by the RAND Corporation, Santa Monica, Calif.
© Copyright 2016 RAND Corporation
**RAND**® is a registered trademark.

*Cover: ninog and ChenPG.*

Support RAND
Make a tax-deductible charitable contribution at
www.rand.org/giving/contribute

www.rand.org

# Preface

Exploiting the rapidly growing sources of data available for collection and analysis is one of the greatest professional challenges facing today's intelligence leaders. The magnitude of potentially relevant data is overwhelming, and more data are being generated and stored every day. Whether the data originate from machines or are based on use of language, the associated analysis makes it possible to uncover important information that would otherwise remain hidden. This type of analysis was impossible only a few years ago, when less data were collected and stored digitally and when information technology systems were incapable of accommodating such large amounts of data.

The question, then, is not *whether* to develop data science capabilities, but rather *how* to do so.

In 2013, the Defense Intelligence Agency (DIA) Directorate for Analysis initiated a program seeking to modernize defense intelligence analysis—specifically, seeking to address the big data problem from the military intelligence perspective and focusing on the inadequacy of existing personnel, tradecraft, and methodologies to manage big data analysis. To address the problem, the Director for Analysis proposed to adopt emerging Intelligence Community–developed tradecraft and methodologies that allow better organization and exploitation of information: object-based production and activity-based intelligence. To address the volume of data becoming available and to learn how to extract more knowledge from these nontraditional data sources, the Director for Analysis identified a requirement within the agency for data science experts. The director asked the RAND Corporation to explore the possibilities of creating a data science capability within DIA that could meet these new demands for the organization's mission and enabling elements. DIA specifically asked RAND to address two questions: What skills do data scientists need and how many does DIA need? And how can data science be organized inside a large organization like DIA?

This report should interest military, defense, and intelligence officials responsible for understanding the implications of data science for future military operations and intelligence activities. Military intelligence officials whose responsibilities include managing the acquisition and use of the flood of data now available to national authorities, allies, and adversaries will be particularly interested. The report aims to provide new insights to intelligence planners, resource managers, and intelligence oversight officers.

This research was sponsored by DIA and conducted within the Intelligence Policy Center of the RAND National Defense Research Institute, a federally funded research and development center sponsored by the Office of the Secretary of Defense, the Joint Staff, the Unified Combatant Commands, the Navy, the Marine Corps, the defense agencies, and the defense Intelligence Community.

For more information on the RAND Intelligence Policy Center, see http://www.rand.org/nsrd/ndri/centers/intel.html or contact the director (contact information is provided on the web page).

# Contents

Preface .................................................................................... iii
Figures and Tables ................................................................. vii
Summary .................................................................................. ix
Acknowledgments ................................................................ xiii
Abbreviations ......................................................................... xv

CHAPTER ONE
Introduction ............................................................................. 1
Study Scope and Structure ........................................................ 3
Methodology .............................................................................. 4

CHAPTER TWO
Data Science Activities in the Private Sector .......................... 5
Improving the Reliability and Quality of Products and Services ............................. 6
Increasing Organizational Efficiency and Agility ..................... 7
Anticipating Threats and Opportunities .................................. 8

CHAPTER THREE
Data Science Education ........................................................... 11
Data Gathering and Organization ........................................... 12
Academic Programs Offer Two Types of Data Science Education ........................... 14
Education Is Diverse and Interdisciplinary, Crosses Different Degrees .................... 15

CHAPTER FOUR
Identifying and Defining Data Science Specialties ................. 17
Characterizing the Four Data Science Specialties ................... 17

CHAPTER FIVE
Data Science Capability in DIA Today .................................... 21
Defining Data Science ............................................................. 22
Current Data Scientists at DIA ................................................ 22

Training Data Scientists.................................................................... 23
Fitting Data Scientists into DIA ........................................................ 23
Perspectives on Data Science ............................................................ 23

**CHAPTER SIX**
**Building and Maintaining a Data Science Capability**..................... 25
Notional Position Descriptions and Grades.......................................... 25
Training Requirements...................................................................... 26
Government Civilians Versus Contractors............................................ 28

**CHAPTER SEVEN**
**Organizing Data Science at DIA**...................................................... 31
Lessons from the Private Sector ........................................................ 31
Variables to Be Considered in Structuring a DIA Data Science Capability ................. 32
Potential Organizational Structures ................................................... 33
Data Science Roundtable.................................................................. 37

**CHAPTER EIGHT**
**Building DIA's Data Science Capability: Findings and Recommendations**............ 41
Key Findings.................................................................................. 41
Recommendations: What DIA Can Do to Create an Enduring Data Science
   Capability ...................................................................................45
The Way Forward ........................................................................... 48

**APPENDIXES**
**A. Interview Protocol** ....................................................................51
**B. Notional Data Science Position Descriptions**...............................55
**C. Methodology**...........................................................................65

**References** ..................................................................................75

# Figures and Tables

## Figures

3.1.   Data Science Curriculum Characterization from Southern Methodologist University's Master's in Data Science Program ..................................... 15

7.1.   Decentralized Structure ............................................................. 34

7.2.   Hub with Coordinating Role ...................................................... 35

7.3.   Hub with Execution Role .......................................................... 36

7.4.   Center of Excellence ................................................................ 37

7.5.   Data Science Roundtable .......................................................... 38

C.1.   Dimensions of Data Science Identified by the Principal Components Analysis ... 70

C.2.   Data Science Terms Depicted as a Network Based on Associations Identified Through Correspondence with Academic Course Descriptions .................. 71

C.3.   Depiction of Data Science Education as a Network of Terms ................... 72

## Tables

6.1.   Manpower Cost-Risk Trade-Offs .................................................. 29

C.1.   Sample Burt Table of Data Science Courses and Terms ......................... 66

# Summary

Technology and the ability to gather and manipulate vast quantities of data have fundamentally altered the way that intelligence organizations collect, process, analyze, and disseminate information. Intelligence leaders worldwide are struggling to understand the implications of the information revolution for intelligence collection and analysis.

The breadth of the data problem, as seen from the intelligence perspective, is striking: The Intelligence Community (IC) needs to collect unstructured and structured data generated by public and private entities, including foreign targets, allied and liaison partners, consumers, and its own organizations. Simultaneously, the IC needs to process and exploit the data collected by an increasingly large and diverse array of sensors—each capturing volumes of data on orders of a far larger magnitude than a generation ago and at multiple levels of analysis and classification—to support a diverse range of consumers, including automated and autonomous control systems, warfighters and military leaders, and elected officials and their appointed agents who are responsible for the nation's strategic decisionmaking.

While some elements of the IC have already established a data science career field, others are in the early stages of exploring the role and function of data scientists and how they can be employed to best support the operations and mission of their particular agency or organization. As part of an all-source intelligence agency, Defense Intelligence Agency (DIA) analysts access a wide array of classified and unclassified information to support national and defense stakeholders, including the military services and combatant commands. Given the unique mix of authorities (e.g., Title 50 and Title 10), requirements, and customers (e.g., national and tactical), the establishment and professionalization of a data science career service within DIA will require tailoring to address the agency's diverse needs and missions.

The DIA Director for Analysis asked the RAND Corporation to identify the key academic and experiential attributes of a data scientist to develop concepts for building a corps of data scientists in DIA and to study workforce recruitment, retention, and training to develop and retain a core data science capability that could support the needs of the entire organization. The goal was for DIA to gain a better understanding of fitting data science into the defense intelligence enterprise, especially all-source analysis; developing data science "product lines"; establishing and managing relations with customers through identifying and addressing technology and database needs; and satisfying the needs of decisionmakers. Each of these issues required research and analysis to maximize the value of data science within DIA and ensure the activities assist and inform DIA's decisionmakers, analysts, and operators.

The research addressed two key questions:

- How should DIA identify and hire data scientists and provide career services for them?
- How should data scientists and activities be organized within DIA?

The study team took a multifaceted approach to address these questions, including a literature review of current commercial applications of data science, a structured analysis of university-level data science education, semistructured interviews with senior analysts and managers across DIA's directorates and mission centers, and direct engagement with DIA's human resource and personnel management leaders and planners, to gain insights into creating and institutionalizing new career services within DIA. The results of this research are organized and presented in the chapters that follow.

In Chapter Two, we examine data science activities in the private sector, identifying three areas in which data science applications have developed: (1) improving the reliability and quality of products and services, (2) increasing organizational efficiency and agility to better respond to changes in the marketplace, and (3) anticipating new threats and opportunities based on competitive trends and risk management.

In Chapter Three, we examine university-level training in data science. RAND's analysis is based on the undergraduate and graduate curricula of 12 universities offering formal degrees, certificates, or academic concentrations in data science and related fields. This examination includes a structured, formal analysis that was used to identify subfields within data science education. We identified four data science specializations: data engineering, data analysis, data communication, and computational social science.

In Chapter Four, we provide a summary of the results of our interviews with DIA managers and analysts. These interviews aimed to identify data science activities already performed within DIA and to capture managers' and analysts' perspectives on opportunities and challenges associated with integrating data science across the enterprise. The interviews revealed that DIA's existing data science activity is unexpectedly large and diverse and that perspectives differed significantly regarding the prospects of data science.

In Chapter Five, we explore ways to recruit and retain government civilian data scientists. We provide notional position descriptions and grades for the data science specialists identified in Chapter Three. We developed these position descriptions with the assistance of DIA's Office of Human Resources. We spoke with an individual at the

Office of Personnel Management to determine whether there was existing documentation on data science position descriptions, but learned that there had been no request to review data science or data science work or to create a series for this type of work. This chapter also explores the opportunities and risks of using contractor data scientists instead of government civilians.

In Chapter Six, we explore four options for constructing a data science capability in DIA. Four organizational structures are characterized, ranging from ad hoc communities of interest and practice to a formal center of excellence that commands organizational resources and personnel. In this chapter, we advocate the formation of a Data Science Roundtable with the explicit purpose of confronting the cultural challenges of organizational reform. This body of senior managers will be collectively responsible for ensuring data science develops within DIA as its leadership and stakeholders demand.

In Chapter Seven, we synthesize the insights from the previous chapters and outline findings and recommendations to inform DIA leadership on the incorporation of data science and data into the organization. The study found that:

- Data science is a team sport. Data scientists work in teams internally, combining different computational, statistical, and other research expertise, and externally, where they collaborate closely with subject-matter experts, decisionmakers, and other stakeholders in their projects and products to ensure the relevance of their efforts. This collaboration is shown to deliver outstanding and often unexpected results.

- It will be difficult to decide which private-sector best practices to import into the IC without first deciding what DIA wants to achieve through the use of data science. The basic question is whether DIA wants to have an organic data science capability or just use the information from activities performed on its behalf. A key element in addressing this question is to consider the range of ways to acquire data science products, ranging from simply purchasing data analytics to conducting all data analytics using in-house experts.

- Given the immaturity of the IC's understanding of how data science can be applied to collection, analysis, and other activities, it is unclear what precise mix of skills, knowledge, and capabilities will be required. This makes it critical to identify the desired output as precisely as possible, whether it be the transactional purchase of data or the development of in-house data analytics. Given these ambiguities, it would be sensible to develop a corps of data scientists, comprising contract personnel, government civilians, and active-duty military personnel, to ensure the broadest representation of technical skills and subject-matter expertise.

- Any type of capability DIA decides to build will incur a recruiting and training requirement. Similar to other areas of technical expertise, data science is quickly evolving, so staff data scientists must have access to training to remain at the cutting edge of capability. Training for the organization's managers and leaders will be critical to embedding data science fully into operations, and any effort to incorporate data science into DIA at the enterprise level should be bolstered by an executive education program.

RAND recommendations are:

- DIA should build a data science capability with a mix of government experts and contractors capable of managing data science activities unique to military intelligence operations. Private-sector experiences in building data science capability and interviews with DIA elements already using data science techniques and data scientists suggest that achieving a quick start will require external expertise, but embedding the tradecraft into the organization will take time and requires a far more-detailed understanding of the needs and culture of the IC.
- DIA should establish a center of excellence or another centralized body to coordinate data science activities; develop tradecraft and methodologies; promote ongoing and completed projects; and advocate for tools, resources, contracting support, and training opportunities for staff members.
- DIA must acquire the right mix of data science experts. A combination of contractors and government civilians/active-duty military will allow DIA to build an agile workforce with capabilities that are fungible across DIA's analytic, collection, and enablement functions.
- DIA will need to establish a training structure for data science. As discussed extensively in Chapter Six, DIA will need to invest in three levels of training to develop an enduring data science capability that will address the needs of the entire agency. The types of training for DIA include an instructional track focused on data science practitioners, a second track for nonspecialists who wish to familiarize themselves with data science tools and concepts, and a third track devoted to executive education on data science applications and managing data science projects and staff.
- DIA should establish a Data Science Roundtable and task its members with responsibility for achieving the internal cultural changes necessary to ensuring data science is an enduring agencywide capability.

# Acknowledgments

Our special thanks go to many RAND experts and colleagues, including Eric Landree, Maria Lytell, Leonard Braverman, and Zev Winkelman. Many thanks to reviewers Ed Burke and Costas Samaras, who provided invaluable feedback in shaping the final product. Barbara Bicksler and Kate Giglio helped us create a coherent oral and graphic narrative for the project's briefing. Our sponsors at DIA—Elmo Wright, MAJ Keith Walthall, and Linwood Creekmore—provided excellent support to the team, and we are indebted to them for making our work easier.

# Abbreviations

| | |
|---|---|
| DEXCOM | Deputy Executive Committee |
| DIA | Defense Intelligence Agency |
| DSR | Data Science Roundtable |
| GENIE | GENetic Imagery Exploitation |
| IC | Intelligence Community |
| INT | intelligence discipline |
| OHR | Office of Human Resources |
| PCA | principal components analysis |
| RDM | Robust Decision Making |

# CHAPTER ONE

# Introduction

The "problem" of big data, or the vast amount of raw information generated across society, is a problem facing every industry and every government in every part of the world. However, this problem is also an opportunity or call to action for DIA analysis. Hidden in this immense volume of data are new information, facts, relationships, indicators and points that either could not be practically discovered in the past or simply did not exist before. We have an opportunity to discover those key pieces of information, possibly prevent strategic surprise or transform all operational aspects of the organization. In order to make these discoveries, we are working to equip our analytic enterprise with the tools, skills and personnel we need to harvest these transformative insights.

—*Cathy Johnston, Defense Intelligence Agency (DIA) Director for Analysis*[1]

Dramatic changes in the international security environment, the proliferation of well-armed state and nonstate adversaries, and technological advances and diffusion present an unparalleled challenge to national, defense, and intelligence leaders. Decisionmakers are being forced to defend against traditional challengers, such as Russia, China, and Iran, while simultaneously dealing with a mix of counterinsurgency and counterterrorism issues elsewhere in the world that involve both new and traditional adversaries. Meanwhile, cyber security threats are demanding increasing attention.

The growth in the complexity of the international environment has been coupled with an unprecedented explosion of data. Technology has made it possible for individuals to create and store data in quantities not imagined even a few years ago. By 2020, the digital universe—the data we create and copy annually—will contain nearly as many digital bits as there are stars in the universe. The digital universe is doubling in size every two years, and by 2020 it will reach 44 zettabytes (44 trillion gigabytes).[2]

---

[1] "Q&A: Cathy Johnston," *Geospatial Intelligence Forum*, Vol. 13, No. 2/3, 2015.

[2] EMC Digital Universe with Research and Analysis by IDC, "The Digital Universe of Opportunities: Rich Data and the Increasing Value of the Internet of Things," web page, April 2014.

Exploiting the rapidly growing sources of data available for collection and analysis is one of the greatest professional crises facing today's intelligence leaders. The magnitude of potentially relevant information is overwhelming, and more is being generated and stored every day. Whether the data originate from machines or human interaction and communication, the associated analysis makes it possible to uncover important facts that would otherwise remain hidden. This type of analysis was impossible only a few years ago, when less information was collected and stored digitally and information technology systems were incapable of accommodating such large amounts of data.

The question, then, is not *whether* to develop a capability to exploit big data, but *how* to do so.

The "personnel" referred to in the opening quote of this chapter are practitioners of a newly minted profession that is developing rapidly in the public and private sectors: data science. Definitions of data science vary, and leading organizations and practitioners note that the field is evolving, having grown out of intellectual and technological roots traced back to the 1960s.[3] Despite the diversity of definitions, there are many commonalities that suggest the existence of a stable, core set of skills and concepts. Specifically, definitions of *data science* emphasize the interdisciplinary research—statistics; computer science, particularly databases; distributed computation; and machine learning—that undergirds big data collection, processing, and analysis. However, other definitions include or emphasize social science, design, business, or computational or quantitative variants of established disciplines, tightly coupling data science activities to applications in business, government, academia, and other fields.[4]

The duties of a data scientist vary across industry sectors, companies, and even within different government organizations. In many cases, the precise duties and expectations of data scientists are still being defined. In the Intelligence Community (IC), some organizations have already established the position of data scientist and have created an associated career field, while others are in the early stages of exploring the role and function of data scientists and how they can be employed to best support the operations and mission of their particular agency or organization.[5]

In 2013, the DIA Directorate for Analysis initiated a program seeking to modernize defense intelligence analysis. The initiative was based on the understanding that from the military intelligence perspective, the big data problem exceeded the existing methodologies and tradecraft of today's defense intelligence professionals. Technology and the ability to collect and manipulate vast quantities of data have fundamentally

---

[3]  Gil Press, "A Very Short History of Data Science," *Forbes*, May 28, 2013; and Michael Hochster, "What Is Data Science? And What Is It Not?" *Quora*, January 16, 2014.

[4]  Vasant Dhar, "Data Science and Prediction," *Communications of the ACM*, Vol. 56, No. 12, December 2013; Jeff Leek, "The Key Word in 'Data Science' Is Not Data, It Is Science," blog post, *Simply Statistics,* December 12, 2013; Berkeley School of Information, "What Is Data Science?" web page, n.d.; and New York University, "What Is Data Science?" web page, n.d.

[5]  For example, see the description for a data scientist position at the Central Intelligence Agency: Central Intelligence Agency, "Careers and Internships: Data Scientist," web page, March 10, 2016.

altered the way intelligence organizations collect, collate, and analyze information, but the development of tradecraft to exploit the information has lagged behind.

To address these problems, the analytic modernization initiative leveraged two processes to allow better organization and exploitation of information: object-based production and activity-based intelligence. These methodologies have developed since the end of the Cold War, but especially during operations in Iraq and Afghanistan, as the intelligence task shifted from a focus on fixed objects to movements of individuals or equipment that could threaten U.S. individuals or interests. The concept behind object-based production is to organize data collected from multiple intelligence disciplines (INTs) around recognizable "objects" to facilitate multi-INT and all-source examination of the information. Activity-based intelligence aims to enable more-timely fusion of this preassembled data into finished, actionable intelligence. Activity-based intelligence is an "analysis methodology that rapidly integrates data from multiple INTs—human intelligence, signals intelligence, imagery intelligence, technical intelligence, and open-source intelligence—and sources around the interactions of people, events, and activities in order to discover relevant patterns, determine and identify change, and characterize those patterns to drive collection and create decision advantage."[6]

To address the volume of data becoming available and to learn how to extract more knowledge from these nontraditional data sources, DIA's Director for Analysis sought assistance from data science experts. To explore the possibilities of using data science methodologies in defense intelligence, DIA turned to the RAND Corporation.

## Study Scope and Structure

DIA asked RAND to address two key issues associated with developing a data science capability: manpower and organization.

The manpower question focuses on how many data scientists are needed, where in the DIA structure they should be employed, what data science position descriptions should look like, and what data scientists' grades/seniority ranks should be.[7]

The organizational question sought insight into how data science should be structured inside DIA, including whether data scientists should be contractors, civilians, active-duty military, or a mix. As an all-source intelligence agency, DIA analysts access a wide array of classified and unclassified information to support national and defense stakeholders, including the military services and combatant commands. Given the unique mix of authorities (e.g., Title 50 and Title 10), requirements, and customers (e.g.,

---

[6]  Chandler P. Atwood, "Activity-Based Intelligence: Revolutionizing Military Intelligence Analysis," *Joint Force Quarterly*, Vol. 77, April 1, 2015. The author cites this quote from an interview he conducted with Jon Kiminau, the U.S. Air Force A2 Analysis Mission Technical Adviser, on October 18, 2013.

[7]  We interpreted "How many data scientists does DIA need?" as a question about the cost of scientists at various levels, from entry to senior, thus allowing DIA to determine the right mix of staff to build the needed capacity in an environment of resource constraints.

military and tactical), it was anticipated that the establishment and professionalization of data scientists for DIA would likely be unique to the agency's needs and missions.

By agreement with the study sponsor, RAND focused on DIA headquarters elements and personnel; DIA personnel at the combatant commands and other military intelligence organizations were not included in the study. However, the findings and recommendations reached may be applicable and scalable to other organizations within the defense intelligence enterprise and elsewhere.

## Methodology

RAND used a mix of methodologies to examine the questions posed by DIA.[8] First, we looked at the private sector, identifying three areas around which data science applications have developed: (1) improving the reliability and quality of products and services, (2) increasing organizational efficiency and agility to better respond to changes in the marketplace, and (3) anticipating new threats and opportunities based on competitive trends and risk management.

We conducted a deep study of academic programs that educate data scientists. This was complemented with research on existing opportunities for data scientists as reflected in job postings. This research allowed us to separate data science expertise into categories, which we believe will help with recruitment and staffing. Chapter Two and Appendix C describe the methodology used in considerable detail.

Additionally, we conducted semistructured interviews with more than 50 DIA managers, analysts, collectors, and enablers across DIA headquarters organizational elements to better understand the current status of data science in DIA, gather perspectives on integrating data science capabilities into DIA activities, and identify what the evolving requirements might be. The results of the interviews are covered in Chapter Five.

Finally, to address position descriptions and potential work levels for data scientists, the study team worked closely with DIA's Office of Human Resources (OHR) to develop notional position descriptions, discussed in Chapter Six and in Appendix C, that could be used to recruit data scientists and place new hires in the organization. The position descriptions are designed to address how many data scientists are needed and what their expertise will cost. We also discussed with the OHR Workforce Analytics team how RAND's work might be deployed to more systematically ascertain the data science capabilities already present in DIA's staff. We also consulted the U.S. Office of Personnel Management to determine whether any focused development of position descriptions or career fields specifically for data science had occurred within that office—none had.

---

[8]   RAND's Human Subjects Protection Committee reviewed the project and methodology and determined the research to be exempt.

# CHAPTER TWO

# Data Science Activities in the Private Sector

RAND was asked by DIA to identify the key experiential attributes of data scientists and to study their recruitment, retention, and training within the workforce. One key is understanding how the private sector employs data scientists and organizes data science capabilities within their organizations.

Data science has emerged as a rapidly growing and increasingly important source of competitive advantage in the commercial sector. Businesses increasingly recognize that their internal data holdings provide unique, exploitable benefits about their corporate strategies, organizational behavior and structure, customers, markets, supply chains, and more. Likewise, increasingly robust, low-latency, and ambient data sources have created opportunities for firms to compete in or transform traditional markets by offering new products and services. The ability to merge private, internally held data with external data sources, whether publicly available or acquired by other commercial services or information brokers, has placed data science and data scientists on the front line of market competition across an ever-expanding frontier of domains.

In the commercial context, data science applications have largely developed around three broad application areas:

- improving the reliability and quality of products and services
- increasing organizational efficiency and agility to better respond to changes in the market and streamline internal processes
- anticipating new threats and opportunities based on the identification of emerging issues, trends, and risk management.

In each case, data science tools and methods have been used differently, revealing the diversity of their utility and potential. However, there are also notable areas or domains in which the contributions of data science—at least in its most common form of collecting and assessing empirical data—have been more limited.

In this chapter, we provide several examples of commercial data science and examine their use in each of the three application areas.

## Improving the Reliability and Quality of Products and Services

Some of the most prominent and promising applications of data science in the commercial sector are found in the interface between firms and their customers. Many compelling examples exist in the domain of predictive analytics, targeted customer engagements, and autonomous, interconnected systems that can improve products and processes.

For example, RAND researchers recently interviewed several commercial research laboratories about their investments in data science[1] and found that one manufacturer of medical imaging systems was using data science to improve medical outcomes for doctors and patients. The company is networking its magnetic resonance imaging machines to vastly increase the quantity of images available to data scientists, who will use the images to train algorithms that assist doctors in making medical diagnoses.

A second example, also drawn from the medical domain, has been the effort by many firms to reduce the rate of patients being readmitted to the hospital within 30 days of their discharge in response to requirements of the Affordable Care Act. Under the law, health care providers must achieve passing scores on several risk-adjusted metrics regarding readmission rates or face several financial, and possibly legal, penalties.[2] Data scientists have worked with health insurers and providers to develop predictive models that estimate the likelihood that a patient will require readmission within 30 days of discharge. Today, many information technology and health care consulting firms are working with health care providers to commercialize and integrate predictive analytic models into the provision of medical services and the administration of patient care during and after hospital visits.[3]

A third example of data science use is in the Rapid Feature Identification Project at Los Alamos National Laboratory. Researchers at the laboratory developed a software package called GENetic Imagery Exploitation (GENIE), which is used to identify features for image analysis.[4] GENIE is a software package that uses genetic algorithms and minimal analyst input to *learn* and identify features in images. This is a job that traditionally would be done by hand by an analyst or researcher. This type of identification can be time-consuming and potentially full of errors. GENIE learns from ana-

---

[1]  Interview with Leonard Braverman, April 2015.

[2]  Centers for Medicare and Medicaid Services, "Readmissions Reduction Program," web page, updated August 4, 2014.

[3]  Examples include IBM's Watson Health, which applied deep machine learning techniques to health care services and administration, and Additive Analytics. See Adrienne Murray, "What Is the Value of Healthcare Dashboards?" web page, HealthCatalyst, August 8, 2013; Bobbi Brown, "A Best Way to Manage a CMS Hospital Readmission Reduction Program," web page, HealthCatalyst, n.d.; "Predicting Hospital Readmissions: Laura Hamilton Interview (Additive Analytics CEO)," *Data Science Weekly*, n.d.; and IBM Software, *Reducing Hospital Readmissions for Congestive Heart Failure*, 2012.

[4]  Los Alamos National Laboratory, "GENetic Imagery Exploitation," web page, 2011.

lysts' input and then automates the feature identification process, minimizing human error, increasing the speed of image processing, and contributing to a higher-quality and more-reliable product.

A final example of data science being employed to improve the reliability and quality of products and services can be found in the emerging Internet of Things, where instrumented and connected devices offer data scientists new opportunities to identify, predict, and address anomalies or failures in a variety of devices, ranging from industrial-scale power plants to personal automobiles and kitchen appliances. Through these networked systems, data scientists have been able to automate system diagnostics and predict the likelihood of specific failure modes, allowing human managers or software control systems to prevent failures or minimize repair times and costs. For example, GE has developed 5,000 turbine electricity generators that have up to 250 embedded sensors that transmit data in real time to a central monitoring facility. When machine readings fall outside prescribed levels, preemptive maintenance is scheduled to minimize downtime and the additional risk posed by larger, cascading failures that may take several weeks to repair.[5]

## Increasing Organizational Efficiency and Agility

In the following examples, commercial organizations used data science to help improve their efficiency and ability to adapt to rapidly changing markets.

The use of predictive analytics to assess current staff and potential hires in competitive markets was popularized in Michael Lewis's book, *Moneyball: The Art of Winning an Unfair Game*, which detailed professional baseball managers' use of advanced metrics to identify the physical potential of players when building a team and evaluating contracts. Data scientists have adopted similar approaches to assessing potential hires by commercial firms to provide a deeper estimate of a candidate's technical capabilities and expertise than can be gained from a few in-person interviews with hiring managers and prospective coworkers. For example, Gild, a company that assesses software engineering and programming talent for clients, employs data science in its evaluations. By examining activities on relevant social and professional software development sites like GitHub and Stack Overflow, Gild's data scientists try to discern the skills and experiences candidates possess that may not be well characterized on their resumes; this lets recruiters create personalized, tailored communications to target prospective employees.[6]

---

[5]  Peter Kelly-Detwiler, "Machine to Machine Connections—the Internet of Things—and Energy," *Forbes*, August 6, 2013; Bala Deshpande, "Connecting Dots: Preventive Maintenance, Big Data, Internet of Things," blog post, Simafore.com, October 16, 2013.

[6]  Gild, "SocialCode + Gild: Using Gild to Discover Talent Who Are 'Hidden Gems' for a Startup with a Unique Culture," web page, n.d.

Data science has been used not only to recruit new employees, but also to manage them—determining important factors like training and development needs, compensation, and performance evaluation and informing decisions about whether and when to promote or terminate employees. By drawing on vast corporate data holdings—employee time cards, expense reports, building accesses, emails, and more—corporate data scientists are able to assess the individual and collective productivity of employees working in increasingly instrumented environments. These assessments go beyond simple metrics of performance, such as billability or absence rates, by seeking to identify internal thought leaders, influence brokers, trust networks among employees, and other hidden structures that are not formally specified by the organization's design.[7]

A third example of data science shaping the internal structure and processes of organizations is changes in corporate logistics. Southwest Airlines developed an agent-based model to examine how their baggage handlers decided the placement of cargo on aircraft. Data scientists interviewed Southwest employees, represented their decision-making processes as algorithms, and simulated them to better understand how baggage handlers' individual choices aggregated to create the organization's actual cargo management system. They were then able to experiment with alternatives for managing cargo and discovered that time and costs could be saved by leaving cargo on a flight for as long as possible, rather than unloading, storing, and reloading it onto an aircraft flying a more direct route. Data scientists recommended new strategies that made the airline more efficient and saved an estimated $10 million over five years.[8]

## Anticipating Threats and Opportunities

Another area of commercial data science applications focuses on assisting commercial organizations with strategic foresight, trend analysis, and executive decisionmaking. These applications are often more complex and tailored than those focused on improving commercial products and services or achieving operational efficiencies. This is due to the need to link data-driven insights garnered from historical and contemporary cases and apply them to future environments under the expectation of changing market conditions. These applications begin to create a common core of agreed defi-

---

[7]   David Karckhardt and Jeffery R. Hanson, "Informal Networks: The Company Behind the Chart," *Harvard Business Review*, Vol. 71, 1993, pp. 104–111; Alex Pentland, "The New Science of Building Great Teams," *Harvard Business Review*, April 2012, pp. 60–70; and Marcus Buckingham and Ashley Goodall, "Reinventing Performance Management," *Harvard Business Review*, April 2015, pp. 40–50.

[8]   Richard Daft, *Management*, Boston: Cengage Learning, 2009, p. 218; Fred Seibel and Chuck Thomas, "Manifest Destiny: Adaptive Cargo Routing at Southwest Airlines," Uncluttered.com, 2000; and Chuck R. Thomas Jr. and Fred Seibel, "Adaptive Cargo Routing at Southwest Airlines," Ernst & Young Center for Business Innovation, 2000.

nitions of data science and approach its fringes, where data science and fields such as knowledge engineering and decision science overlap.

One application of data science has been the development and use of prediction markets. Prediction markets are a type of distributed computation in which the aggregated, independent assessments of many individuals often predict discrete outcomes with greater accuracy than individual experts.[9] Hollywood movie studios have employed prediction markets to forecast box office returns of movies before their release, and Google and Microsoft have used prediction markets to estimate the demand for new products and services and the completion time and release of complex software development projects. In these cases, data scientists work with data generated by market participants responding to specific statements about the future. By buying or selling shares in a specific event occurring, market dynamics can determine a price that corresponds to a probabilistic estimate. From this information, decisionmakers can be made aware of trends, emerging threats, opportunities, and shifting market conditions.

Sentiment analysis is another data science application used by commercial organizations to monitor the popularity of their products and services, track the strength of their brands, and detect strategic moments to enter or exit a market.[10] Sentiment analysis, like prediction markets, aggregates the opinions of populations, but does so by starting from unstructured data—predominantly text but also imagery, video, and audio—to assess public sentiment involving products, brands, and experiences. Sentiment analysis supports corporate marketing and strategic communication campaigns by helping decisionmakers assess whether their products, services, and brand name are becoming more or less embedded in targeted populations.

The development of Robust Decision Making (RDM) tools that seek to identify strategic choices that perform well in the face of uncertainty is another area where data science has been used to identify threats and opportunities. RDM rests on the forefront of data science applications because of its dual emphasis on data analysis and data generation via computer simulation. While most data science applications emphasize the development of predictive models, RDM is more concerned with how models can be used during deep uncertainty when decisionmakers and analysts cannot agree on

---

[9]  James Suroweicki, *The Wisdom of Crowds*, New York: Anchor, 2005; Jared M. Schrieber, *The Application of Prediction Markets to Business*, thesis, Cambridge, Mass.: Massachusetts Institute of Technology, June 2004; Kenneth J. Arrow, Robert Forsythe, Michael Gorham, Robert Hahn, Robin Hanson, John O. Ledyard, Saul Levmore, Robert Litan, Paul Milgrom, Forrest D. Nelson, George R. Neumann, Marco Ottaviani, Thomas C. Schelling, Robert J. Shiller, Vernon L. Smith, Erik Snowberg, Cass R. Sunstein, Paul C. Tetlock, Philip E. Tetlock, Hal R. Varian, Justin Wolfers, and Eric Zitzewitz, "The Promise of Prediction Markets," *Science*, Vol. 320, No. 5878, May 16, 2008, pp. 877–878; and Daniel O'Leary, "Prediction Market as a Forecasting Tool," *Advances in Business and Management Forecasting*, Vol. 8, 2011, pp. 169–184.

[10]  Jenna Dutcher, "Sentiment Analysis Symposium: Uncovering Human Motivations," blog post, *DataScience@Berkeley*, March 7, 2014.

a framework for assessing problems or ranking outcomes.[11] By emphasizing the use of multiple alternative models combined into ensembles, data scientists can help decision-makers find strategies that perform well across several alternative frameworks based on different evaluative criteria. As a result, RDM allows data scientists to provide an analytic framework that helps multiple, often competing stakeholders agree on a course of action.[12]

RDM has been applied in many domains, both within commercial organizations and across broad communities of interest where commercial industries, government regulators, and the public interact, such as in the maintenance of watersheds and basins that are at risk of depletion. By developing computational models that emphasize interactions within complex systems—such as demand for water resources, allocation rules, climate and hydrology, and current and future technological capabilities—participating stakeholders can find actions that suggest good outcomes in all frameworks. RDM also helps them identify where perspectives diverge, such as disagreements about future demand or how commercial farmers and manufacturers may adapt to alternative regulatory regimes or changes in tax policy. What distinguishes RDM from other data science activities is that it is specifically tailored to support decision-making under uncertainty through the systematic generation of data, often petabytes or more, to illuminate complex interactions, key uncertainties, and challenging scenarios that would require actors to change their behavior.[13]

Examples from the private sector will certainly inform the development and organization of data science activities in intelligence organizations, but circumstances in the IC and DIA are distinctly different. From a practical standpoint, the IC faces organizational and operational constraints that many private-sector firms do not. IC data are collected and classified in many ways, do not have a standardized structure, are not uniformly accessible in a single system, and often have limitations on how they can be used. IC employees are generally civil servants, and they must be cleared to handle classified information. Moreover, government manpower structures are less flexible than those in the private sector, so manning and structuring a data science capability will necessarily take a different direction.

---

[11] Robert J. Lempert, Steven W. Popper, and Steven C. Bankes, *Shaping the Next One-Hundred Years: New Approaches to Long-Term Policy Analysis*, Santa Monica, Calif.: RAND Corporation, MR-1626-RPC, 2004.

[12] RAND Corporation, "Robust Decision Making," web page, n.d.

[13] Riddhi Singh, Patrick M. Reed, and Klaus Keller, "Many-Objective Robust Decision Making for Managing an Ecosystem with a Deeply Uncertain Threshold Response," *Ecology and Society*, Vol. 20, No. 3, 2014; Matteo Giuliani, Andrea Castelletti, Francesca Pianosi, Emanuele Mason, and Patrick M. Reed, "Curses, Tradeoffs, and Scalable Management: Advancing Evolutionary Multi-Objective Direct Policy Search to Improve Water Reservoir Operations," *ASCE Journal of Water Resources Planning and Management*, Vol. 142, No. 2, 2014; Joseph R. Kasprzyk, Patrick M. Reed, and David M. Hadka, "Battling Arrow's Paradox to Discover Robust Water Management Alternatives," *ASCE Journal of Water Resources Planning and Management*, Vol. 142, No. 2, 2014; and David M. Hadka, Jonathan Herman, Patrick Reed, and Klaus Keller, "OpenMORDM: An Open Source Framework for Many-Objective Robust Decision Making," *Environmental Modeling & Software*, Vol. 74, 2014.

# CHAPTER THREE

# Data Science Education

DIA asked RAND to review key academic attributes of a data scientist to facilitate DIA's potential recruitment of data scientists and to identify possible sources of external training for its existing personnel. To provide DIA with a perspective on data science education, we analyzed courses of instruction for undergraduate and graduate degree programs from 12 universities offering degrees or certifications in data science. Further analysis done at DIA's request provided insights into academic subfields within data science by identifying clusters of topics, techniques, and methods found in academic data science instruction.[1] The analysis allowed RAND to separate data science expertise into four distinct categories, which helped identify the types of expertise needed to build a complete data science capability. This analysis is covered extensively in Chapter Four and in Appendix C.

Understanding academic data science education is important for two reasons as DIA considers how to develop an internal data science capability. The first is a practical matter: By identifying graduates' likely exposure to data science concepts and tools, DIA's senior leadership can set expectations regarding the capabilities that prospective applicants coming from different academic programs may possess and the roles they can play within the organization. These expectations may be tailored to specific types of data science programs or even extended to additional degree programs that provide relevant training under other titles (e.g., bioinformatics or computational journalism). Given the diversity of definitions of data science and differences in academic programs, DIA managers should be aware that people may be called data scientists, but have different skills to match the needs and expectations in their specific field.

The second, more subtle, reason is that the structure of university data science education provides a window into the world of data science careers by revealing stable and well-defined specialties. Chris Wiggins, a prominent data scientist at the *New York Times* and Columbia University, noted that once formal degree programs at universities are stood up, it indicates that new professional specializations have largely reached

---

[1]  This assessment did not consider courses from professional data science training and education services offered through professional associations, nonuniversity-associated online programs, etc.

a settling point and that stable job categories exist in the marketplace, allowing academic programs to prepare graduates for these defined job positions and careers.[2]

## Data Gathering and Organization

To identify the structure of data science education, course descriptions from 21 degree programs offered by 12 universities were compared with a dictionary of 4,000 data science and related terms. This approach let us create a high-dimensional correspondence matrix in which key data science terms, their synonyms, and their acronyms were associated with one another through the ways in which academic courses were described. The objective was not to identify the frequency of different topics or concepts in data science education, but rather to assess the diversity of data science education and identify additional academic disciplines whose instructional curriculum may be consistent with data science concepts and methods but go by other names.[3]

The 20 degree programs that were examined in our analysis were:

- Carnegie Mellon University, Master of Information Systems Management, concentration in Business Intelligence and Data Analytics
- DePaul University, Master of Science in Predictive Analytics
- George Mason University, Department of Computational and Data Sciences, Doctorate in Computational Social Science[4]
- Georgetown University, Department of Biochemistry and Molecular & Cellular Biology, Bioinformatics Graduate Program
- Harvard University, Extension School, Data Science Certificate
- Indiana University, School of Informatics and Computing, Master of Science in Data Science
- New York University, Center for Data Science, Master of Science in Data Science
- Northwestern University, McCormick School of Engineering, Master of Science in Analytics
- Northwestern University, School of Professional Studies, Advanced Data Science Certificate Program
- Northwestern University, School of Professional Studies, Advanced Medical Informatics Program

---

[2]   Sebastian Gutierrez, *Data Scientists at Work*, New York: Apress, 2014, p. 14.

[3]   The programs used in this analysis do not include all data science programs that have ties to the IC or are prominent within the field. Instead, they were selected because, collectively, they covered a wide array of data science offerings.

[4]   In fall 2015, the Department of Computational Social Science in the Krasnow Institute for Advanced Study became the Program in Computational Social Science in the Department of Computational and Data Sciences.

- Northwestern University, School of Professional Studies, Analytics and Business Intelligence for IT Professionals Advanced Graduate Certificate Program
- Northwestern University, School of Professional Studies, Database & Information Technologies Certificate Program
- Northwestern University, School of Professional Studies, Online Master's in Predictive Analytics
- University of Rochester, Goergen Institute for Data Science, Bachelor of Arts in Data Science
- University of Rochester, Goergen Institute for Data Science, Bachelor of Science in Data Science
- University of Rochester, Goergen Institute for Data Science, Master of Science in Data Science
- Southern Methodist University, Dedman College of Humanities & Sciences, Applied Master of Science in Statistics and Data Analytics Program
- Southern Methodist University, Online Master of Science in Data Science
- Stanford University, Institute for Computational & Mathematical Engineering, Master of Science Program, Data Science Track
- Syracuse University, Newhouse School, Master's in Computational Journalism.

The selection of these programs was based on a combination of factors, such as their ties to notable data science practitioners, online availability of course descriptions, diversity of offerings in data science education, and novelty, in order to serve as exemplars of relevant academic training that might not have appeared in research on data science education already performed by DIA. Note that the selection of specific programs in this analysis was not intended to endorse or promote the programs examined over others that were not included in this dataset.

The structured analysis of these programs was based on course listings and descriptions available online. In several cases, descriptions for the same course varied based on instructor and/or term offered. In these cases, the more-recent offering of the course was preferred unless an older description provided greater detail as to the instructional goals and content of the course. Likewise, different academic programs at the same school often taught courses with the same name in different programs. In these cases, both versions were included in our dataset.

RAND researchers developed a data science dictionary by taking terms and concepts from several disciplines and aggregating them into two files that were used in the analysis.[5] The first file was a single list of terms that represented concepts, methods,

---

[5]    The dictionary of terms was developed from the following sources: DataFloq, "An Extensive Glossary of Big Data Terminology," web page, n.d.; DataInformed, "Analytics and Big Data Glossary," web page, updated September 24, 2014; AnalyticBridge, "Data Science Dictionary," blog post, November 17, 2012; Harish Kotadia, "Key Big Data Terms You Should Know," blog post, April 9, 2013; Statistics.com, "Glossary of Statistical Terms," web page, n.d.; Kirk Borne, "Big Data A to ZZ—A Glossary of my Favorite Data Science Things," Converge

and tools. The second was a set of acronyms and synonyms associated with key terms, which allowed us to account for differences in the use of acronyms, spelling, punctuation, and discipline-specific terms that represented the same scientific or technical concept. For example, "Agent-Based Modeling" is a simulation method used to study complex systems; in the context of academic course descriptions, it might appear as "Agent Based Modeling," "Agent-Based Models," "ABM," or "Individual-Based Modeling," as it is known in ecology.[6] Note that certain terms could not be resolved within the bounds of the analysis performed in this study. For instance, "Individual-Based Modeling" is often referred to as "IBM," but when combined with computer science terminology or academic curriculum, "IBM" might refer to the Fortune 500 company.

After combining all the terms from these dictionaries into a single list, a secondary list was developed that included associating acronyms and synonyms. This list served as the original basis of the structured analysis and was then manually adjusted based on how well it covered the academic course descriptions. For example, in statistics, the term "course" has a specific technical meaning regarding the treatment of a sample population. However, in the context of analyzing academic programs, the term "course" is overrepresented and loses its underlying scientific meaning, as phrases like "this course examines" can be found in the beginning of most course descriptions. As a result, this term, as well as others, such as "class" from computer science, was subsequently removed from the dictionary. The resulting dictionary contained over 4,000 terms.

## Academic Programs Offer Two Types of Data Science Education

There are two types of academic data science programs. First, general programs emphasize broad exposure to concepts, methods, and tools that data science professionals should be aware of and familiar with prior to entering the workforce. Second, more specialized, applied programs exist that are oriented toward specific career fields, such as business, finance, medicine, genetics, or journalism. In these cases, educational offerings mix general data science tools and methods with substantive, domain-specific theory, data sources, and research methods. The structure of data science as

---

Blog, March 21, 2014; KDnuggets, "Data Mining and Predictive Analytics Glossary," blog post, n.d.; Jeffrey Goldstein, "Resource Guide and Glossary for Nonlinear/Complex Systems Terms," PlexusInstitute.org, n.d.; ComplexityBlog.com, "Glossary of Terms," blog post, n.d.; Shelly Palmer, "Data Science 101: Definitions You Need to Know," blog post, September 7, 2014; Data Analytics and R, "What Is . . . " blog post, n.d.; and various authors, Quantitative Applications in the Social Sciences, series, Thousand Oaks, Calif.: SAGE Publications, 1987–2014.

[6]   See Volker Grimm and Steven F. Railsback, *Individual-Based Modeling and Ecology*, Princeton, N.J.: Princeton University Press, 2005; Steven F. Railsback and Volker Grimm, *Agent-Based and Individual-Based Modeling: A Practical Introduction*, Princeton, N.J.: Princeton University Press, 2012; and Uri Wilensky and William Rand, *An Introduction to Agent-Based Modeling: Modeling Natural, Social, and Engineered Complex Systems with NetLogo*, Cambridge, Mass.: MIT Press, 2015.

a career field should be largely discernable from the offerings of general data science academic programs due to their efforts to prepare students for careers in a wide variety of professional fields or future academic studies, rather than discipline-focused programs that prepare students for entry into specific professions. Based on this broad background, the structure of data science, as identified by the principal components analysis (PCA), revealed unexpected insights into data science as compared with the descriptions offered by general data science education programs.

General data science programs have tended to emphasize three tracks, or specialties, that all students must be familiar with: computer science, statistics, and data visualization. For example, Southern Methodist University's master's in data science program depicts its curriculum as a combination of the three (Figure 3.1).

## Education Is Diverse and Interdisciplinary, Crosses Different Degrees

A cursory examination of the 20 programs listed above reveals several aspects of data science education. First, there is significant diversity in the types of data science education available; programs cover education at the undergraduate and graduate levels and in resident and nonresident/online programs. This list does not include other options, such as nonuniversity programs like Coursera or Udacity that offer training in several computational and mathematical topics and tools, and intensive, specialized programs offering training to researchers outside their formal academic programs, such as the Santa Fe Institute's Complex Systems Summer School and Inter-University Consor-

**Figure 3.1**
**Data Science Curriculum Characterization**
**from Southern Methodologist University's**
**Master's in Data Science Program**

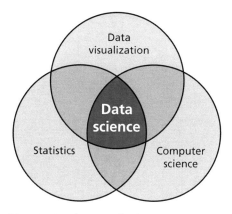

SOURCE: Southern Methodist University,
"Data Science @ SMU," web page, undated.
RAND *RR1582-3.1*

tium for Political and Social Research's Summer Program in Quantitative Methods of Social Research.[7]

Second, the data science programs listed above are interdisciplinary in nature, founded in relatively new academic disciplines such as biochemistry, research institutes, or professionally oriented and applied programs. These programs largely exist in traditional science, technology, engineering, and mathematics departments and schools, but they also can be found in social and life sciences schools and departments, business schools, and journalism schools. An interesting point regarding the diversity of data science education can be seen in the University of Rochester's undergraduate degree in data science, which offers students the option to pursue a bachelor of arts or a bachelor of science degree.

Finally, the rapid development of data science programs was made possible by its interdisciplinary roots and the ability to draw on academic institutions' existing faculty and courses. As a result, many courses that data science students take are cross-listed with other programs. Indeed, the formation and rosters of data science research institutes that house many academic programs reveal the diversity of data science research and education, covering a range of topics, including highly technical and mathematical problems in computer and software engineering and statistics as well as consulting skills and the ethics of data access and manipulation, artificial intelligence, and automation.

---

[7]   See Coursera.org; Udacity.com; Santa Fe Institute, "Complex Systems Summer School," web page, n.d.; and Inter-University Consortium for Political and Social Research, "Summer Program in Quantitative Methods of Social Research," web page, n.d.

# CHAPTER FOUR

# Identifying and Defining Data Science Specialties

RAND was asked by DIA to study workforce recruitment, retention, and training implications for a data science capability. Our analysis—performed with the standard tools of data science—of more than 600 courses offered by data science programs at 12 major universities found unexpected patterns in the course offerings that allowed us to identify "clusters" of expertise. The analysis also helped to identify data science specializations for DIA and the development of potential job categories that would comprise a comprehensive data science capability and career service at DIA. We concluded that four distinct specialties are required for a complete data science capability: data engineer (computer science), data analyst (statistics), data communications specialist (data visualization), and computational social scientist. We note that the standard big data analysis tools we used both aided the analysis and demonstrated how data scientists can derive unseen knowledge from previously unexploited data sets. An extensive discussion of the analytic methodologies used can be found in Appendix C.

## Characterizing the Four Data Science Specialties

To address DIA's question on how many experts were needed and at what work levels, we worked with in-house RAND human resources experts and the OHR at DIA to develop notional position descriptions for each of the identified areas. The position descriptions are intended to identify work levels and salary ranges for data scientists recruited externally or internally. For each of the specialties, we defined specific job requirements. We developed four notional positions, one for each specialty and one for each work level: entry, full, senior, and expert. Position descriptions for all work levels will need to be developed for the positions. The tracks are not mutually exclusive, but overlap.

**Data Engineer.** The data engineer is a technical adviser, working with multiple types of databases on matters related to capturing and processing live, streaming, and distributed data. This specialist designs and develops customized data collection, management, and search-and-retrieval systems to support the collection, processing, exploitation, analysis, and dissemination of big and complex datasets. The data engineer would also be responsible for supporting the acquisition and development of enterprise-level computational resources and customized tools for use by individual or small groups of analysts, collectors, and enabling staff members. Data engineers focus on the development of software systems but work closely with the chief information officer's (CIO's) office to articulate requirements for specialized hardware and other infrastructure needs.

**Data Analyst.** The data analyst is responsible for providing descriptive statistics, probability models, and other quantitative assessments of raw, processed, and generated data. This specialist uses a combination of traditional statistical and machine learning/artificial intelligence techniques to analyze big and/or complex data sets in support of analytic, collection, and managerial activities. Data analysts work with a variety of databases and types, and are familiar with multiple quantitative and computational methodologies. The data analyst creates tailored analyses for a variety of different users in support of collection, analysis, and managerial activities and decisionmaking.

**Data Communications Specialist.** The data communications specialist is responsible for communicating and presenting summaries of structured and unstructured data in visual, text-based, and interactive formats to nonspecialists. This requires strong written, oral, and graphical communication skills. This specialist develops static visualizations and graphics of complex and big data sets, produces dynamic dashboards for monitoring analytic measures and discovering patterns and anomalies in data, provides analytic narratives and stories to explain the meaning and implications of quantitative analysis, and provides transparency into complex analytic processes by explaining how quantitative and computational methodologies gather and manipulate data in support of analytic, collection, and managerial activities. This position needs an artistic mind to conceptualize, design, and develop reusable graphic/data visualizations that viewers may interact with and that can be dynamically updated with new information. This position also requires very strong technical knowledge to implement such data visualizations using the very latest technologies.

**Computational Social Scientist.** The computational social scientist is a technical adviser, translating social science theory into computational algorithms and developing prototype models for simulating social behavior and processes and analyzing data on social structures and interactions. Computational social scientists provide methodological support to regional and functional analysts, assist in the development and application of quantitative and computational analytic methods, produce finished analysis based on quantitative and computational methods, and support decisionmakers through the generation and evaluation of tailored data sets.

PCA, the methodology described in Appendix C, identified these three subfields or specializations within data science. The first two, identified by their opposite scores in the PCA results, were the central and familiar distinction between computer science and statistics. This result matches both the intuition of practitioners and the formal descriptions of data science offered by academic programs.

Another component of data science was identified through the PCA when analyzed with different variables. Specifically, this component consisted of a mixture of organizational, psychological, and cognitive sciences and has not been explicitly identified in the formal descriptions of data science academic programs. This third component was identified based on projections of the data into the second and third dimensions, which identified organizational theory (inclusive of synonyms such as organizational behavior, organizational science, etc.) and a branch of terms strongly associated with cognitive science, psychology, linguistics, and artificial intelligence. We binned these dimensions of data science into a broad subfield called computational social science, for historical and practical reasons.

In theoretical and historical terms, the structure, processes, and dynamics of organizations have been a subject of study in all social science disciplines, and the notion of organization and collective behavior and interaction are the *sine qua non* of social behavior and systems themselves.[1] Core social science concepts—such as rationality, utility, authority, power, reciprocity, identity, norms, institutions, strategic interaction, and communication—offer perspectives on the ways in which individuals cooperate, collaborate, and compete with one another in order to identify and achieve individual and collective goals.[2] Cognitive science, neuroscience, and artificial intelligence emerged as critical components of the social sciences with the invention of the computer because it provided new opportunities to simulate the processes by which individuals and organizations came to perceive their environments, make choices, plan, and implement actions.[3]

Computational social science was termed as such because of the prominence that human and organizational behavior have in its conceptual composition. Indeed, every social science has provided important studies of organizational behavior, cognition, and decisionmaking, even extending into the biological sciences and other system sciences in which organisms and organizations are equated with one another.[4] Moreover, the emphasis on organizational behavior, cognition, decisionmaking, and artificial intelligence also corresponds with data science's applications to many domains that emphasize supporting decisionmakers at operational and executive levels of business and government organizations, as demonstrated by several business intelligence and business analytics programs.

---

[1]   Hunter Heyck, *Age of System: Understanding the Development of Modern Social Science*, Baltimore, Md.: Johns Hopkins University Press, 2015.

[2]   Jon Elster, *Explaining Social Behavior: More Nuts and Bolts for the Social Sciences*, New York: Cambridge University Press, 2007; Detrich Reuschemeyer, *Usable Theory: Analytic Tools for Social and Political Research*, Princeton, N.J.: Princeton University Press, 2009; Gary Goertz, *Social Science Concepts: A User's Guide*, Princeton, N.J.: Princeton University Press, 2012.

[3]   Samir Okasha, *Philosophy of Science: A Very Short Introduction*, New York: Oxford University Press, 2002, pp. 10–11.

[4]   Heyck, 2015.

By comparison, visualization was not identified by the PCA as a distinctive cluster or specialization of data science. This is not to suggest that visualization is not a relevant skill set for data science; indeed, a manual inspection of the curricula of academic programs reinforces its importance in data science education and practice. Instead, it indicates that visualization, and communication skills more broadly, are equally associated with other specializations and receive emphasis in most academic programs regardless of their other differences.

Several observations about this categorization of expertise should be emphasized. First, while entry-level individuals may have expertise in only one area, expertise in the other areas should be developed over time, and that expertise should be coupled with professional intelligence expertise. Data science experts will become more valuable as they learn more about the intelligence profession. (Similarly, current intelligence professionals will need to acquire some level of expertise in data science and data analytics to improve their professional capabilities.) Second, it is worth noting that a data science capability needs to have each of the four areas represented in some way; while DIA may need more computational social scientists, a full data science capability will require expertise in all four areas. Our research on private-sector data science activities and interviews with DIA staff emphasized that data science is a team sport.

Two obvious questions are how to recruit these individuals and whether it is wiser to retrain existing intelligence professionals through data science certification programs, rather than hire data scientists and give them intelligence training. In either scenario, it will be critical to educate the analytic, collection, and management units so that they understand how to engage in a dialogue that identifies specific needs with greater specificity. Just as data scientists need to become familiar with—but not experts in—intelligence activities, intelligence officers need to become familiar with the skills and capabilities data scientists have that can potentially contribute to improving intelligence collection and production. Enhancing the mutual understanding of capabilities, through regular interaction or by use of pilot projects, will help identify best practices for effective collaboration.

# CHAPTER FIVE

# Data Science Capability in DIA Today

To better understand the context in which DIA might develop a data science capability, we sought to determine the extent to which data science is currently employed across the agency and what the evolving requirements for incorporating data science might be. We conducted approximately 25 semistructured interviews with more than 50 managers, analysts, collectors, and enablers across DIA. These included individual interviews and group sessions with some of the regional and functional centers. In addition to interviews, we participated in a regular meeting of a data science community of interest (primarily made up of DIA Directorate of Intelligence staff) to gain an initial understanding for who is practicing data science already and what some of the perceived barriers to data science expansion at DIA might be.

The research team leveraged the expertise of RAND's Survey Research Group to develop a standard interview protocol that was then tailored as appropriate for meetings with managers and analysts. The protocol can be found in Appendix A.

The protocol was intended to elicit information in five broad research areas. First, we asked a series of questions to see which techniques or methods the interviewee thought were part of data science. This discussion not only helped us understand how different parts of DIA view data science but also helped us shape how this project defines data science. Second, we asked how data science is incorporated into DIA as an intelligence agency and into the interviewees' respective offices. We intended this questioning to characterize what data science looks like in DIA today in terms of the extent of its use, types of missions, types of techniques, and availability of data sets. Third, we covered what opportunities might exist within DIA to receive training. These questions were designed to determine how prevalent training opportunities and requirements were for current employees and what might be required in the future. Fourth, we discussed interviewees' views on how data science should fit within DIA's workforce architecture. These questions were particularly timely, as DIA was in the process of transitioning to a new civilian personnel system made up of specific career fields, which provided an opportunity to consider the appropriate space for those doing data science work. Finally, we asked questions regarding how data science is perceived at DIA and in the IC writ large to help identify any cultural challenges to the incorporation and growth of a data science capability.

In the interviews, we received feedback that clarified the context in which data science developments in DIA will occur. The discussions were rich and varied and produced interesting ideas, including some contradictory views and recommendations. Although data science capabilities are employed in many DIA offices, these capabilities are often not called data science. Not surprisingly, those who use data science techniques now—the early adopters—are strong supporters. Similarly, those who do not use the techniques currently are not strong supporters and are more likely to question how the techniques could be used in their line of analytic work. There also remains a fair bit of skepticism about how data science will be utilized and institutionalized within the agency.

Throughout the course of our interviews at DIA, five themes emerged from the five general question areas: (1) data science definition, (2) current data scientists at DIA, (3) training, (4) fitting data scientists into DIA, and (5) DIA's perspective on data science.

## Defining Data Science

Views varied widely at DIA on the definition of data science. Individuals' definitions were clearly linked to the output of their organizations, so different organizations described data scientists and data science activities differently. Some focused on algorithmic or computational research methods and decision support. Others focused narrowly on big data, computer science, and statistics. In the mission services area, the focus was on using data analytics to realize enterprise operations efficiencies and improvement.

## Current Data Scientists at DIA

From the interviews, we learned that the organizations in DIA are using data science techniques, but individuals doing the work are not labeled data scientists. This may be in part because some offices use other titles, such as "statisticians" or "computer scientists," but it could also be because practitioners of data science specialties are distributed across the agency and have limited organized professional contact and no specific career field or other human resources structure around which to organize. Nonetheless, we did identify individuals who served functions similar to those of a data scientist. In one case, a contractor was serving this function on a critical project but then left the contracting company, leaving the office with a need to fill that role. This office decided to invest in the cadre capability by hiring a DIA civilian "statistician" to perform the function. Another officer we interviewed sought an individual with a computer science and math background to perform a specific technical analytic function. Although a centralized team of what one might call data scientists exists within the modernization initiative at Analytic Enterprise Operations, many interviewees were either unaware of the group or uncertain how to use the group's ability to assist with their individual missions.

## Training Data Scientists

Interviews generated little information regarding training or training requirements for personnel conducting data science at DIA. Those we interviewed from various DIA offices did not provide or require such training at this juncture. However, pockets exist where organizations internally or externally train their personnel in data science. Where it does happen, offices fund their personnel to attend one-off training events focused on data science techniques, and a few individuals mentioned that they are pursuing training on their own time to build their expertise in data science techniques. Several noted that data science techniques constantly evolve, requiring continuous education and on-the-job training to learn and practice new methods. Any effort to institutionalize data science into DIA will need to consider the substantial investment in training.

## Fitting Data Scientists into DIA

Perspectives about how and where data scientists should be used in DIA varied significantly. First, interviewees disagreed on who should serve as data scientists for the agency. Some saw data science as an empowering tool that analysts could and should use, and they advocated for training current analysts to also perform this task. Others argued that data science is a professional field, and not all analysts—who are currently trained to write reports but whose skills will likely be expanded in the future as analytic output grows to include more graphic content—will be able to learn the appropriate skills; they believed a data science specialist was needed to supplement analytic capability.

Views also varied on where data scientists would "sit." Some advocated for a centralized hub at the enterprise level that would allow for more efficient use of resources and could also assist with senior decisionmaking. The rationale for a centralized hub was that data science tools are adaptive and could be applied to any mission. Others, including those with experience working closely with data scientists, argued that data scientists would have greater impact if they were embedded with analysts on a team, so that they work every day on the same mission. They also were concerned about the scarcity of data science resources in a centralized organization, predicting that the "threat of the day" would get priority. Between these two ends of the spectrum, several interviewees gravitated toward a hub-and-spoke model, in which data scientists would be managed by a core element but assigned to individual offices to better understand and support their missions.

## Perspectives on Data Science

As mentioned, perspectives on data science within DIA might be summarized most appropriately by what one interviewee called the "haves" versus the "have nots." Those who use data science techniques now are big supporters; they have generally seen improved value in their products. Those who have not used data science techniques express less support or may question how the techniques could be used in their line of work. In this latter group are those who fear data science will fundamentally change or eliminate their job.

From all perspectives, interviewees concurred that data science will not succeed at DIA if it does not have support up the chain and at the highest levels. Many perceived managers as reluctant to invest in data science, with several citing a previous data science effort that helped to save the agency a considerable amount of money but was eliminated anyway due to a lack of leadership support. A common practical concern among managers was where the billets would come from for data scientists. Managers who supported the concept of data science were quick to say that they would not hire a data scientist into an existing billet or recode an existing billet for data scientists. They expected the agency to provide a billet for the activity to signal that the agency's senior leadership endorsed and valued data science.

Current practitioners of data science within DIA emphasized that, because previous efforts failed to take root, a cultural change would be imperative to institutionalize what the capability can contribute to the agency writ large.

CHAPTER SIX
# Building and Maintaining a Data Science Capability

A key issue for DIA leaders and managers is the cost of building a data science capability. In an era of flat or declining budgets, managers continuously struggle to balance the need to retain current staff and capabilities while identifying and acquiring new talent to meet new tasks. Developing a new capability such as data science raised several concerns: competition with other entities also seeking top-flight talent; security clearability of data scientists; availability of trained data scientists at midcareer level; and the immediate and long-term costs (personnel, training, tools) of building the capability. To further complicate the question, the effort to construct a data science capability is being considered while DIA restructures its civilian personnel system from rank-in-position to rank-in-person. This restructuring has required establishing career fields to provide for professional growth for the agency's intelligence professionals. After identifying the precise skills that would be needed to build a data science capability (Chapter Four), the study team decided to build standard civil service position descriptions for the various data science experts to facilitate cost estimates and provide insight into how many data science experts might be needed.

## Notional Position Descriptions and Grades

To determine the number and types of experts needed and at what work levels, the researchers worked with in-house RAND personnel subject-matter experts and OHR staff at DIA to develop notional positions descriptions for the four identified emerging data science specialties (data engineer, data analyst, data communications specialist, and computational social scientist) described in Chapter Four. For each, RAND identified specific IC work responsibilities, job requirements, and the potential training/certifications desired, and then aligned each specialty with a DIA work level: entry, full, senior, and expert (notional position descriptions can be found in Appendix B).

Once the notional position descriptions were drafted, we sought additional validation of the job requirements and desired training/certifications by comparing the descriptions to similar positions in the commercial sector. Using the four position titles as search terms on the job website LinkedIn.com, we found over 100 similar position descriptions that allowed us to review the most prevalent and valued skills, education, and experience that companies sought for each position as well as the typical responsibilities and expectations of the four specialty tracks.

The proposed specialty tracks are not mutually exclusive, but overlap. While entry-level individuals may have expertise in only one area, expertise in the other areas likely would and should be developed over time as data science responsibilities are executed. That expertise will be coupled with professional intelligence expertise that will also grow over time. Most importantly, an effective data science capability needs each of the specialty areas represented in some way. While DIA may need more computational social scientists to meet its needs than a commercial data science capability, a full and effective data science capability at DIA will require expertise in all these areas.

## Training Requirements

### Training for Data Science Practitioners

Whether DIA hires data scientists from the outside or retrains existing intelligence professionals by providing data science credentials, additional training will be required. Training devoted to data science practitioners within the organization should emphasize two key features. First, a training program should seek to broaden data scientists' range of exposure to the other data science specializations or subfields. In the commercial sector, data scientists are encouraged to change jobs frequently to increase their exposure to different technologies and concepts, and to expand their network. Intrinsic value is placed on the breadth of types of problems they have worked on and what tools and methods they are familiar with. Given that government employment does not generally experience or support such rapid turnover, data scientists within the organization should be given opportunities to develop projects and collaborate across the broader organization to create virtual experiences that expand their technical and methodological breadth. Data scientists should be encouraged to gain exposure and proficiency in all specializations, even if they are working primarily in just one or two.

Second, a training program for data scientists should ensure that government data scientists do not lose touch with their commercial and academic peers. In other domains, maintaining currency has often proved difficult as staff members and their managers often prefer to interact with other IC insiders to mitigate security risks and avoid inadvertent revelations of classified material. Given the pace at which data science is developing, such a fraying of networks can result in the organization and its operational units quickly falling far behind their commercial and academic counterparts, leaving DIA unable to fully perform its responsibilities to the best degree possible. Thus, data scientists within DIA need the support of a robust outreach program (discussed in greater detail below) specifically tailored to allow them to stay in contact and keep pace with their peers outside the IC.

## Data Science Training for Non–Data Scientists

The training of non–data scientists, particularly at the middle manager level, in data science tools and concepts is necessary to provide DIA with three organizational capabilities. First, it will increase the quantitative skills and expertise of the staff broadly defined. Second, exposing nonspecialists to data science, even at a rudimentary level, can reduce the intimidation associated with data science tools and methods and provide nonspecialists with a set of concepts and a vocabulary for interacting with specialists. Just as staff members in traditional line analyst positions are not expected to be analytic methodologists, training in analytic methodology nevertheless increases the rigor and transparency of their products and raises awareness as to when specialists may be needed to assist with research or production. Finally, exposing DIA staff members to data science training can help identify those with an aptitude and interest in data science, which may be important and necessary if external hiring constraints are imposed on DIA.

## Data Science Training for Executives

Executives within DIA should receive training on data science tools and concepts, with a focus on raising awareness and expectations of what possessing a corporate data science capability means for DIA and critical cultural factors that promote or inhibit the full utilization of data science within the organization. Given that data scientists will provide data and analysis to senior managers, senior managers must learn how to be responsible taskers and consumers of these projects and products.

## Professional Qualification Maintenance and Enhancement

Developing and maintaining a data science capability is a dynamic process that requires a sustained commitment to training to keep pace with a rapidly developing body of technologies, concepts, and application areas. DIA will need to develop a strategy that accounts for not only recruiting data scientists but also maintaining and growing their skills and expertise over time.

Data scientists should have an organizational commitment to ensuring their access to academic and professional training opportunities, particularly continuing education. Many of these trainings and certifications can be attained online, but staff members should be encouraged to attend training programs in person when appropriate to develop their personal networks. Data scientists could also take advantage of the collaboration and networking opportunities available on the IC's classified systems, including specialist conferences and publishing in classified journals.

Additionally, DIA may consider establishing a robust data science outreach program and capitalize on the depth and breadth of talent in the Washington, D.C., region. A data science outreach program may include working with local universities, nonprofits, and other organizations to create tailored educational programs for DIA staff and executives, sabbatical and Intergovernmental Personnel Act opportunities for faculty members to serve as embedded data scientists within DIA on unclassified or classified programs, sponsored conferences to create opportunities for staff members to meet peers and colleagues in academia, and rotational assignments to allow DIA staff to spend extended time at universities, local labs, and businesses learning about data science applications and research that may be relevant to DIA's middle- and long-term needs.

## Government Civilians Versus Contractors

Given the immaturity of the IC's understanding of how data science can be applied to collection, analysis, and other activities, it is unclear what precise mix of skills, knowledge, and capabilities is required. Because of these uncertainties, organizations are taking different approaches to developing data analytic capabilities. The possibilities include buying analysis, data renting versus owning, contract services, and in-house hiring of data scientists for analysis of in-house data. There are pros and cons to each approach.

The most limited level of ownership is the simple purchase of data science analysis. This is a one-time buy for a one-time provision of data. The advantages to this approach are that it is timely and costs are negotiable. The major disadvantages are uncertainty about the data and data quality, no access to the process, and no access to the experts. Without insight into data sources or the actual databases, assessing the information's validity is impossible beyond taking the company's reputation for good analysis as an assurance.

Contracting for data sets and data science services, rather than simply purchasing the analysis, would create a slightly more intimate relationship with the contractor providing the analysis. This still does not give access to the algorithms, data, or experts, but it does maintain a more significant relationship with the provider. This relationship can help build trust and understanding for future services and contracts. This will be more costly than simply buying data but less costly than owning the analysis capability. However, it involves a higher level of risk because there is no internal capability to quality control the data or services provided.

Developing a service provision capability will also require careful contract negotiation; data science techniques and methodologies are developing quickly, so any contract vehicle will need to be flexible enough to evolve with the profession and to ensure the ability to tap into new expertise as the situation requires.

A slightly different alternative to this arrangement would be to use contract data science experts to analyze internally "owned" organization data. While the organization's level of trust in its data might be higher, the ability to oversee the quality of the data science analytics would still be limited due to the absence of in-house data science expertise. It could be particularly problematic for the military intelligence community to transfer internally owned, sensitive data to an external contractor. One way around the external transfer of data is to host a security-vetted contractor. Both physical and virtual proximity to the contracted data scientists would provide a higher level of control. This method allows all forms of data sources to be brought in and transferred to the contracted data scientist. This means the algorithms, data, and processed data are all visible to and controllable by the data owner. Major downsides to this approach are finding appropriate candidates with the technical skills and subject-matter expertise and vetting them in a timely manner. Another potential issue is ensuring that data scientists have both the broad range of skills necessary to perform and the ability to keep up with the quick evolution of new techniques.

The most costly but most reliable capability would be full-time civilian government data scientists. A full-time data scientist could connect with subject-matter experts and develop the skills and knowledge to provide agile, trusted support. In-house data scientists have controllable tasking and schedules. The data scientist can develop relationships and come to understand specific fields because of the likelihood of longer-term tenure in the position. The downsides are up-front recruiting, training, and long-term costs. Another potential problem is career growth; having a small number of data science experts means potential limits to their professional growth and their transition to more senior assignments or fields. Finally, the lure of higher salaries in the public sector cannot be dismissed.

Table 6.1 summarizes the advantages and disadvantages of various options for acquiring data science services or personnel.

Using active-duty military personnel to fill data science positions is also a possibility. A single example of the use of active-duty military personnel was identified in the U.S. military intelligence community, where a U.S. Army officer trained as an operational research/systems analysis expert has been assigned to help manage a small data science effort.

**Table 6.1**
**Manpower Cost-Risk Trade-Offs**

| Employment Option | Advantages | Disadvantages |
| --- | --- | --- |
| Buy data science analysis | Immediately available; buy what you need and can afford | Uncertain data quality; no access to algorithms; no control over experts |
| Contract for data and data science services | Recurring availability; focus on specific issues | Uncertain data quality; no access to algorithms; no control over experts |
| Contract for services, provide data | Immediate availability; enhance quality and amount of data | No access to algorithms; hard to use with sensitive data; no control over experts |
| Hire contractor data scientists | Continuous availability, known skills; trusted; cleared; no training | Cost, relationships to subject-matter experts; trust (e.g., Snowden) |
| Hire government data scientists | Available; trusted; taskable | Long-term cost; need to train; less subject-matter agility |

# CHAPTER SEVEN

# Organizing Data Science at DIA

Working for an all-source intelligence agency, DIA analysts access a wide array of classified and unclassified information and use a variety of sophisticated analytic techniques and methodologies to support national and defense stakeholders, including the military services and combatant commands. Given the unique mix of authorities, requirements, and customers, establishing a data scientist capability for DIA is only the first step in the process. A second and equally complex task is to integrate that capability into enterprise-level operations.

Building this new capability into the organization without significant new funding—when the agency is already operating near capacity—is an even more complex undertaking that involves decisions on structure, manpower, money, and authority. That said, agency leaders who were interviewed for the study recognized that building a strong capability involves more than just hiring data scientists. To attract the best and brightest individuals, DIA will need a structure that both enables data scientists to apply their expertise to the most complex intelligence problems and includes a human resources process that ensures these new experts can grow as professionals.

These are difficult questions. First, as discussed in Chapter Five, there is already a large but dispersed data science effort under way in DIA. The efforts are decentralized and specific to the organizations in which they are taking place. Restructuring an existing capability that was established for reasons unique to the time and purpose is surely a recipe for contention.

Successful establishment of a robust data science structure in the agency will involve carefully balancing these variables. In the following paragraphs, various organizational options are explored and their benefits and risks highlighted.

## Lessons from the Private Sector

In Chapter Two, we discussed the roles and organizational structures of data science in the private sector. Data science is an enabler for organizations; it is project-focused and often uses data gathered and controlled by the organization. However, the organization of the data science activity is increasingly a concern. One data scientist noted that "Applied data science is all about putting the people who drill the data in constant touch with those who understand the applications. In spite of the mythology surrounding geniuses who produce brilliance in splendid isolation, smart people really do need each other. Mutual stimulation and support are critical to the creative process, and science, in any form, is a restlessly creative exercise."[1]

A growing body of professional literature is examining new structures for data science. Some organizations and organizational theorists have recognized the need to realize the full value of data science expertise by making data science an integral part of the organization. The literature suggests that integration means support from the highest organizational levels, having an identified specialized sector leadership that participates in management, and having career growth opportunities. In some cases, organizations, such as the *New York Times*, have established chief data scientist positions to oversee the application of expertise to their operations. In other cases, large businesses are using a center of excellence concept to structure their data science efforts; Intel's chief data scientist runs the Data Science Center of Excellence, a weekly and voluntary internal forum at Intel designed to enable data scientists within the company to help each other tackle big problems in their respective groups.[2]

Examples from the private sector will certainly inform the development and organization of data science activities in intelligence organizations, but circumstances in the IC and DIA are distinct. IC data is collected and classified in many ways, does not have a standardized structure, is not uniformly accessible in a single system, and often has limitations on how it can be used. IC employees are generally civil servants who must be cleared to handle classified information. Moreover, government manpower structures are less flexible than those in the private sector, so manning and structuring a data science capability will necessarily take a different direction.

## Variables to Be Considered in Structuring a DIA Data Science Capability

Data science is an enabler for most organizations and appears to be structured and subordinated to meet immediate organizational goals. In the business world, data science is project-focused and often uses data gathered and controlled by the organization. This is not the case in the IC, where the data are often unstructured, are not owned by a single organization, or are tightly controlled and unavailable to all analysts.

To explore the complexities of building a data science structure in DIA, the study team identified through interviews with DIA managers a set of basic variables that

---

[1]   James Kobielus, "Data Scientists: Grow and Sustain a Center of Excellence," blog post, *IBM Big Data & Analytics Hub*, May 21, 2012.

[2]   Jessica Davis, "Intel Chief Data Scientist Shares Secrets to Successful Projects," *Information Week*, November 16, 2015.

will need to be addressed by any structuring scheme: responsibilities, authorities, manpower, and organizational placement.

*Responsibility* is defined as the tasks the organization is expected to fulfill. Interviews with those in DIA elements that already have a data science capability show that data scientists collaborate closely with individual subject-matter experts to address complex problems or to exploit large data sets for specific information.

*Authority* is defined as the ability to direct individuals to take action.

*Manpower* is defined as the number of people required and available to perform a specific function.

*Organizational placement* is defined as location in the hierarchy, which determines how the roles, authorities, responsibilities, and manpower are assigned, controlled, and coordinated, and how information flows among the different levels of management.

## Potential Organizational Structures

In this section, we explore four different options for organizing data science in DIA, considering the implication of each organizational structure for the key variables. For each structure, we have identified the organizational location and subordination and briefly discussed manpower. We kept in mind the charter for this study: There are no new manpower resources to build a new organization, so structures will have to be developed within existing manpower ceilings.

### Decentralized (Status Quo)

The decentralized model (see Figure 7.1) is what currently exists at DIA. Data science activities are staffed and resourced as local initiatives, and there is little to no connection between and among them. There is an informal Data Science Community of Interest, organized by the Directorate for Analysis, which represents a bottom-up effort to share best practices and ideas. However, it has no official authorities or ownership of resources, and participation is strictly voluntary. Interviews conducted throughout the agency indicated that few elements outside the Directorate for Analysis were aware of the organization.

This structure today provides a basic level of data science capability to DIA. In its present state, the experts provide an immediate, tailored service that responds precisely to local requirements. The specialists are funded locally and comprise a mixture of government and contractor personnel.

In this structure, data science experts have no special career field and have limited career mobility and professional growth opportunities; all growth and professional enhancement opportunities depend completely on local managers. Managers and experts satisfied with these arrangements have little incentive to change the culture or to look more broadly to determine how their expertise can be scaled to a larger organizational level.

**Figure 7.1**
**Decentralized Structure**

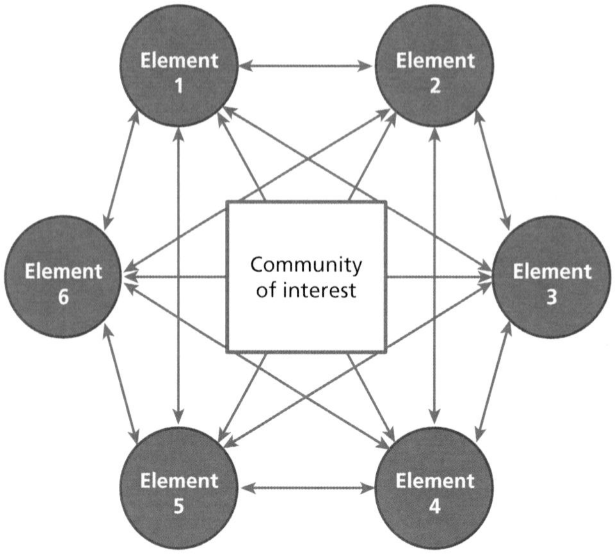

RAND *RR1582-7.1*

## Hub with Coordinating Role

In this hub-and-spoke model (see Figure 7.2), the hub would be responsible for coordinating data science activities across the organization and advocating for data science capabilities to senior levels of the organization. In this model, the hub would have limited authority to make decisions or allocate assets. It would not manage data science manpower directly, but it could influence the use of that expertise through development of time-limited cooperative projects. All resources for data science would remain with the organizations that currently own/fund the activity.

A coordinating hub would have considerably greater visibility at the enterprise level than the decentralized structure. It would provide an internal coordination capability that does not exist in the decentralized scheme. A hub could be tasked and held accountable for advocating for data science expertise, tools, and training at the enterprise level, though the organizational placement of the hub would have a major impact on its ability to be a successful advocate. Finally, a hub could provide a single entity that could be held accountable by DIA's senior leadership for data science issues and policy advice.

On the downside, the hub could be seen as another boutique organization favored and funded (at the expense of others) by current leaders. Depending on the relationship between the hub and its partners in the spokes, an "us versus them" mentality could develop without significant efforts to manage the expectations of both parties. As with all such arrangements, there is the persistent possibility of confusion in authority, so a strong concept of operation would need to be written to avoid this possibility.

**Figure 7.2**
**Hub with Coordinating Role**

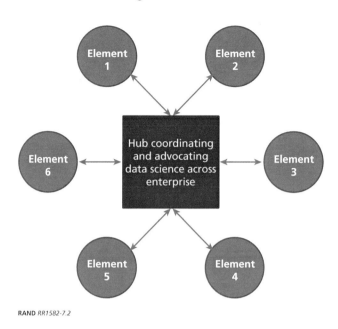

A hub with a coordinating role would not be able to provide services—a shortcoming that could compromise its reputation and ability to function as an advocate for resources and policies. Finally, a small hub could be unable to achieve the critical mass and gravitas required to succeed without a strong leader and strong, visible leadership endorsement.

**Hub with Execution Role**

The third model (see Figure 7.3) envisions a stronger hub-and-spoke model, with the authority and manpower to provide data scientists and data science services to organizational elements. In this model, some data scientists would work in the hub while others would remain embedded in their existing organizations. (Deciding who moves and who stays would likely be contentious.) The hub would deploy data scientists to or embed them within local units. The hub would retain executive and managerial functions, but data science activities would still occur at local levels by data scientists working for the hub.

A strong hub would have agencywide visibility and be in a strong position to serve as an advocate for expertise, tools, methodologies, and tradecraft. It would provide agile, agencywide access to services through applying expertise and developing fungible experts who could fulfill data science support roles in multiple elements of the agency. This hub would be visibly accountable to DIA senior leadership. Similar to the simpler, less-capable hub, this approach could also be seen as another "boutique" organization. Establishing authority and acquiring manpower would be a challenge

**Figure 7.3**
**Hub with Execution Role**

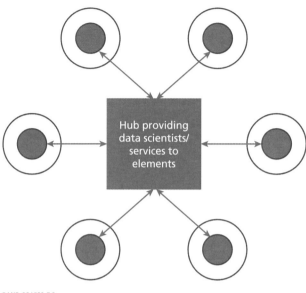

Hub providing
data scientists/
services to
elements

because of the transfer of manpower and the negative perceptions of those elements losing manpower to another. There could be conflict between providing services and building a cross-agency capability if the demand for services outstripped the plan to develop a broader capability.

## Center of Excellence

A final model (see Figure 7.4) would be to centralize all data science activities in the hub. Existing data science experts would be reassigned to the hub, which would centrally manage data science activities throughout the organization. Locating all data scientists within a single Center for Excellence would make the Center responsible for providing all data science services to agency elements. The Center would have the authority to make decisions about allocating resources and assets to address agencywide problems. The Center would own all data science manpower. Data scientists would not be reassigned but would remain in the hub to maximize their contact with one another and to allow Center leaders to build strong expert teams with specific capabilities relevant to the problems to be addressed.

The Center would be situated at a senior level sufficient to be visible to the workforce and have both the visibility and reality of support by DIA leadership. It would be accountable to DIA senior leaders through representation at the various DIA management forums where key policy, operational, and resource decisions are made. The Center would provide strong advocacy for data science policies, tools, training, and

**Figure 7.4**
**Center of Excellence**

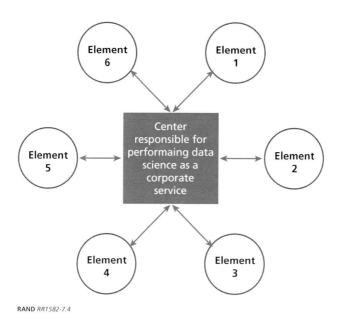

application. To avoid the appearance of creating another boutique organization, the Center could be authorized for a fixed (potentially three-year) term that could be renewed based on its success on performance metrics established at the outset.

## Data Science Roundtable

As noted, selecting any of the models above will result in advantages and disadvantages. The more-decentralized structures will have (or be seen to have) responsibilities that they may not have the resources or expertise to fulfill. Failure to demonstrate data science's value could risk compromising the utility of data science and thus risk leadership and resource support. Establishing a more centralized capability that involved reassignment of manpower would create tensions with organizations losing manpower and money. Assignment of responsibilities and authorities could be complicated if authorities previously executed elsewhere were reassigned. Given the immaturity of data science techniques and methodologies in intelligence today, establishing a strongly centralized capability would not necessarily guarantee successful application of data science techniques. Finally, none of the models really addresses changing the culture to ensure that data science becomes a permanent part of tradecraft.

To address these conflicts, RAND consulted organizational experts to try to identify an innovative approach to lessen potential internal agency conflicts over building a new capability that would require sharing of existing authorities, responsibilities, and manpower. The structure that evolved from these discussions was named the "Data Science Roundtable" (DSR) to distinguish it from other, more familiar bureaucratic names and structures (see Figure 7.5). The DSR would likely benefit any of the four options described above.

The DSR would be composed of senior executives nominated by their directorate leaders. It would be composed not of data scientists, but of professional managers at the Defense Intelligence Senior Level/Senior Executive Service rank. They would meet as a collaborative group to discuss and make recommendations on enterprise-level data science policies and resources. The DSR would report directly to the Deputy Executive Committee (DEXCOM). Recommendations would be based on consensus among the members. If consensus cannot be achieved, the recommendation could be elevated to the DEXCOM.

Conceptually, the DSR is a collective decisionmaking body. Establishing a DSR and giving it the authority and responsibility to recommend decisions avoids some of the contentious issues of the structures described above. In the case of the DSR, no authorities or resources move. Elements with a stake in data science decisions would have the opportunity to identify a senior representative to serve on the DSR to represent specific interests. Existing data science activities would be undisturbed, and

**Figure 7.5**
**Data Science Roundtable**

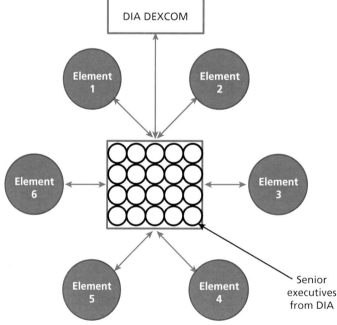

issues concerning potential new data science activities, especially those requiring cross-agency expertise, could be decided at the DSR level.

The DSR would create an environment that is friendlier to data science and allow the activity to develop and mature within DIA while minimizing contention over division of responsibilities, resources, and authorities. Over time, the DSR could be the leader for organizational change and learning. Similar to the Center of Excellence described above, the DSR could be established for a limited time, with its possible extension based on reaching predetermined measures of effectiveness.

The DSR, via its subordination to the DEXCOM, would have strong, visible leadership support. Through the senior staff assigned to represent their elements on the DSR, the organization would possess executive authorities to recommend allocation of resources and manpower. It would fully represent all agency elements and their requirements. Based on DIA leadership directions, it could be held responsible for affecting organizational cultural change.

Lack of data science expertise could be problematic for this leadership organization. Establishing a chief data scientist position in DIA and assigning the person to the DSR would ameliorate this issue. This collaborative organization would lack strong internal leadership, which would place a great premium on building constructive relationships. Despite the well-known difficulties in changing cultures, the DSR would be at least an intermediate step in transforming the DIA organization and culture into one that appreciates the expertise required to successfully manipulate large data pools to extract new knowledge.

# CHAPTER EIGHT

# Building DIA's Data Science Capability: Findings and Recommendations

We examined private-sector data science activities and data science education and training courses to answer DIA's questions concerning the identification and recruitment of data science expertise and options for organizing data science within DIA. Analyzing the data gathered through data science tools, we identified academic and experiential attributes of data scientists. Specifically, we isolated four unique areas of expertise in which data scientists are being trained. This insight allowed us to develop notional position descriptions and associated grade levels that provided some insight into the experts needed and the potential costs to DIA of building a corps of data scientists.

Separately, using information from interviews with DIA leaders and managers, we identified several concerns that would need to be taken into account in building an agency-wide capability. Drawing on relevant best practices from the private sector, using the insights developed from close study of the academic sector, and incorporating the context gathered from the interviews, we identified several characteristics of an effective data science environment and developed a set of recommendations that DIA might consider implementing to build and sustain a strong, enduring data science capability for the entire enterprise.

## Key Findings

**Data science is a team sport.** Collaborative work is the essence of data science research and production. Data scientists work in teams internally, combining different computational, statistical, and other research expertise, and externally, collaborating closely with subject-matter experts, decisionmakers, and other stakeholders in their projects and products to ensure the relevance of their efforts.[1] This collaboration is shown to deliver outstanding and often unexpected results. Similarly, interviews with current practitioners of data science in DIA also highlighted the criticality of collaboration. The Office in Infrastructure Analysis provided a number of examples of collabora-

---

[1] D. J. Patil, "The Importance of Taking Chances and Giving Back," in Carl Shan, Henry Wang, William Chen, and Max Song, eds., *The Data Science Handbook*, Middletown, Del.: The Data Science Bookshelf, 2015, p. 19.

tion that led to discovering previously unknown information. Another example is the departing statistician in one of the regional centers; the capability the departing contractor provided was seen as so critical to the overall analytic output of the office that a government billet was allocated to hire a full-time government civilian who could provide and expand the capability.

An unexpected discovery of data science "teaming" emerged from the analysis of data science education, which showed that data science professionals are educated and trained for different specialties. We believe that these areas of specialization will need to be represented on the data science team; several data science specialists may need to collaborate with subject-matter experts to achieve a high-quality outcome. The differences in expertise among data scientists emphasize the need to clearly identify the expected outcome of incorporating data science into the intelligence process so that the right expert is assigned to the problem. This reinforces the idea that the data science "team" needs to incorporate all areas of expertise, whether the team is all contract workers, all government workers, or a mix.

For example, elements within DIA expressed different needs and desires for data science within their geographic, functional, or operational areas: Where one unit may benefit from identifying trends and patterns in volumes of data that overwhelm individual inspection and curation, others needed to identify microscopic inconsistencies in data elements that may reside on multiple systems to resolve individual identities and attribute activities in space and time. In each case, data science may make important contributions to their respective missions, yet the particular skills, tools, and techniques would vary in order to be relevant.

Analysts we interviewed frequently mentioned another aspect of teaming: data sharing. Exploitation of classified data will require cleared experts, whether they are contractors, full-time civilians, or active-duty military. Accessing classified data collected and owned by different agencies and stored in separate databases will be challenging and will likely require approval by the organizations responsible for the information. Introducing data analysis tools into classified systems will need to be approved and managed. Processes and policies that do not exist today will need to be developed to ensure that the data and insights emerging from the analysis of the data are handled ethically and legally.

**It will be difficult to decide which private-sector best practices to import into the IC without first deciding what DIA wants to achieve through the use of data science. The basic question is whether DIA wants to have an organic data science capability or just use information from activities performed on its behalf.**

Deciding how to invest in data science—that is, in data, analysis tools, and personnel—is challenging. As our research has shown, understanding how to use data science to improve operations and efficiencies in the private sector has not been quick. Leaders sometimes made risky decisions to invest in capabilities that seemed promising. For military intelligence practitioners, this investment will be hard because the promise of results is uncertain and the costs may be larger than expected.

In the private sector, the use of data science, even when used experimentally and for research and exploratory analysis and production, occurs with clearly defined expectations and commitments regarding the relationship among data science, data science teams, and the larger organization. As Chapter Two discusses, some organizations exploit data science for their own business or operational purposes; some businesses provide data aggregation and/or analytic services; and others provide expertise to accomplish the exploitation of big data for other entities, where the data are proprietary or classified or otherwise restricted. Some consulting services have data scientists for hire. In each case, data science activities are staffed with the data science expertise necessary to achieve the outcome. The central question to be addressed is whether to purchase analysis, use contract experts to build a data science capability, or develop the capability using government civilians (or a mix of contractor and civilian personnel). If the answer is "some of everything," then decisions will have to be made balancing the long-term investment in internal capability with the short-term purchase of data and/or services. Importantly, government organizations must also determine whether desired applications of data science would involve inherently governmental activities, which would prohibit opportunities to rely on external support or services.

**Given the immaturity of the IC's understanding of how data science can be applied to collection, analysis, and other activities, it is unclear what precise mix of skills, knowledge, and capabilities will be required.** The uncertainty of the outcome of the application of data science to intelligence, coupled with the plethora of sources, analytic priorities, and technologies and methods, will make it difficult to decide which kinds of expertise and tools to pursue. This makes it critical to identify as precisely as possible what output is desired, whether it be the transactional purchase of data or the development of in-house data analytics. Given these ambiguities, it would be sensible to develop a corps of data scientists composed of contract personnel, government civilians, and active-duty military personnel to ensure the broadest representation of technical skills and subject-matter expertise. Identifying the expected, specific products of data science will be difficult, but dividing the task into specific segments will assist in setting expectations for the organization regarding investments in data science capabilities that are reasonable and reflective of DIA's mission needs, support, and director's vision.

*Purchase of data science analysis.* The most limited level of ownership is the simple purchase of data science analysis. This transactional activity can provide quick results and may be beneficial if existing providers of data and analytic services already provide the reporting that DIA desires. This approach does not build long-term relationships or produce tailored products unique to DIA's specific needs, but it incurs no unspecified or additional costs for data or labor beyond those directly involved in the transaction. The major disadvantages to this are uncertainty in the data and data quality, no access to the process, and no access to the experts. Without insight into data sources or the actual databases, assessment of the validity of the information is impossible beyond taking the company's reputation for good analysis as an assurance. Moreover, if DIA's only data science capabilities existed via transactions of this type, it would lack the internal resources to independently validate the quality of what was provided and be a smart consumer of marketed services.

*Contractors.* The second level would be to have a contractor provide data science experts and/or expertise using either contractor-provided or government data. In this arrangement, contracted experts would provide whatever data science service is requested. This would offer neither access to the algorithms and proprietary tools nor any insight into the quality of the resulting analysis. Developing this kind of relationship will also require careful contract negotiation. Data science techniques and methodologies are developing quickly, so any contract vehicle will need to be flexible enough to evolve with the profession and to ensure the ability to tap into new expertise and technologies as the situation requires.

*Full-time civilians.* The third level of capability foresees hiring full-time civilian government data scientists. A full-time data scientist could connect with subject-matter experts and develop the skills and knowledge to provide agile, trusted support. This means that the data scientist is in-house, so tasking and schedule are controlled. This is useful because the data scientist can develop relationships and come to understand specific fields because of the likelihood of longer-term tenure in the position. The downsides are the up-front recruiting, training, and long-term costs. Another potential problem is career growth; having a small number of data science experts who provide critical skills means potential limits to professional growth and transition to other more senior assignments or fields. Finally, the lure of higher salaries offered by the private sector cannot be dismissed—particularly as opportunities and activities that were previously only performed by the government are increasingly performed commercially, e.g., working with high-resolution remote sensing data.

There is a substantial data science capability in DIA today, although it lacks cohesion and direction. Analysts conducting or benefiting from the use of data science analytics value it because it expands their knowledge and sharpens their analysis. Analysts and managers unfamiliar with data science are wary of its application because they believe it will either increase their workload or eliminate analytic positions. Some are wary of the quality of information generated through data analytics. Managers, even those who are data science enthusiasts, are wary of building data science capability using local resources because the leadership has not yet endorsed the techniques.

**Any type of capability DIA decides to build will incur a recruiting and training requirement.** The application of data analytics to defense, military, and intelligence functions requires expertise that is generally not in abundance in today's military or intelligence organizations. At present, major U.S. intelligence organizations are approaching the task differently. One proposal foresees recruiting data scientists and pairing them with subject-matter experts across the organization. The goal would be to explore how data science techniques can be harnessed to support current and future operations. Another model foresees using centrally organized data scientists who would provide services to elements as necessary. Both structures would recruit from external sources—whether contractors or government personnel—and thus would have to compete with private and commercial organizations that also seek qualified experts. In most cases, commercial employers offer greater compensation and fewer personal restrictions.

Whether the data science capability is composed of contractors, government civilians, active-duty military, or a mix, it will be critical for hiring officials to understand and assess the backgrounds of data scientists, academic data science curricula, and data science training and certification courses. Chapter Two provides an extensive description of the academic programs and the types of expertise currently being produced at universities. Using this list as a basis for recruiting data scientists, whether contractors or government, will provide a degree of insight into the capabilities of the individuals and some assurance that the capability sought is being acquired.

Similar to other areas of technical expertise, data science is quickly evolving, so ensuring that data scientists on staff remain at the cutting edge of capability means they will need access to training, seminars, and other outreach activities to keep DIA staff in contact with data science experts so that their skills and professional networks continue to keep pace.

Finally, training for the organization's mangers and leaders will be critical to embedding data science fully into operations, and any effort to incorporate data science into DIA at the enterprise level should be bolstered by an executive education program. There should be a menu of courses that provides complementary enhancement and sustainment training to analysts and managers on joint production concepts, vehicles, and tools.

## Recommendations: What DIA Can Do to Create an Enduring Data Science Capability

**Recommendation One: DIA should build a data science capability with a mix of government experts and contractors who are capable of managing data science activities unique to military intelligence operations.** Prior to making a decision concerning acquisition of expertise or organizational structures, DIA must decide whether it wants to build an organic data science capability or simply use the information resulting from activities performed on its behalf. Whatever data science structure DIA decides to create should immediately tackle this question. Based on our interviews, it is unlikely that seeking data science products or expertise from only the private sector will be sufficient. Data science practices in the private sector can be imported into the IC, but the scale of the IC problem, the potential critical uses of the results of data science analysis, and the quality of and access to intelligence data available to data scientists and analysts will need to be studied carefully because the IC data science tradecraft will be different. The answers to these questions will shape the data science structure that is established and help determine the types of expertise required.

**Recommendation Two: DIA should establish a Center of Excellence, or another centralized body, that will be responsible for coordinating data science activities, developing tradecraft and methodologies, promoting ongoing and completed projects, and advocating for tools, resources, contracting support, and training opportunities for staff members.** A Center of Excellence provides the best way to quickly insert data science into operations, but cultural change across the agency will still be necessary to ensure the new methods and tradecraft become routine. Considering the broad array of activities already under way in the agency, a quick start is desirable so that common practices can be developed and implemented and so that data scientists across the agency develop communications links that will allow the exchange of expertise. Using contract experts, carefully selected to ensure the right mix of expertise identified in Chapter Four, will allow DIA to move quickly. But long-term dependence on contractor experts will not lead to the kind of agile, enduring data science capability tailored specifically to military intelligence needs that DIA wishes to build, and it will not create a cadre of intelligence officers with deep expertise in data science.

**Recommendation Three: DIA must acquire a robust mix of data science experts.** DIA must consider two factors of this recommendation: the type of data science expertise it acquires and the sources where this expertise can be acquired.

As explained in Chapter Four, data science comprises four distinct subsets of expertise. DIA's data science community will need to ensure all these areas of expertise are represented; the precise mix of skills will be best determined by initially identifying what advantage DIA expects to achieve from data science and the types of analytic or collection work to which data science can best be applied. This task may not be a difficult one initially; individuals interviewed for this report have already thought about how and where data scientists can initially be used.

Once DIA decides where to invest in data science, the second question is how to staff the positions. A combination of contractors and government civilians/active-duty military will allow DIA to build an agile workforce with capabilities that are fungible across DIA's analytic, collection, and enablement functions. The immaturity of data science in intelligence makes it difficult to determine precisely what mix is needed, so investing wisely in a diverse set of capabilities is important for enabling effective experimentation and learning within the organization.

Staff members and contractors hired as data scientists should demonstrate expertise in one or more of the core data science specialties. Usually, this expertise would be evident based on academic achievements and prior professional, military, or other experience. As noted in Chapter Three, there are a large number of data science programs and relevant areas of research and education at the university level. Given that these programs are dynamic, as is data science as an academic and professional discipline, it would be imprudent to view only individuals coming from named data science programs as capable of being data scientists within DIA. Prominent data scientists can be found in physics, chemistry, engineering, political science, journalism, and other fields, so application evaluators should focus on candidates' specific concentrations, coursework, projects, and other training.

DIA should consider canvassing existing staff to identify individuals who are currently applying data science–like techniques in their work or who have an interest in participating in training or certification programs to become data scientists. The number of individuals currently using data science techniques is large. Training/certifying these individuals as data scientists could accelerate integration of data science into operations because these retrained individuals would already have intelligence officers' professional skills. Identification of these individuals could be easily accomplished through either existing agency communications means (notification of training, announcement in internal communications documents, etc.) or by applying the dictionary of data science terms developed for this research to the DIA personnel information database.

Because data science education is rapidly evolving, many practitioners may lack a traditional university education. In instances when potential hires may lack formal university training, many employers have moved to practical exercises and assessments of an applicant's portfolio of project work. In these cases, organizations have given prospective applicants an examination in which candidates download an unclassified dataset and perform several standard operations on the data as well as generate original analysis, visualizations, and other insights that can be evaluated by prospective managers, peers, and human resource officers.

**Recommendation Four: DIA will need to establish a training structure for data science.** As discussed extensively in Chapter Six, DIA will need to invest in three levels of training to develop an enduring data science capability that will address the needs of the entire agency. The types of training for DIA include an instructional track for data science practitioners; a second track for nonspecialists, especially midlevel managers of analysts and collectors who wish to familiarize themselves with data science tools and concepts; and a third track to educate executives on data science applications and managing data science projects and staff. Each of these three levels is intended to facilitate increased integration of data science expertise and applications within the organization and targets different types of providers and consumers of these capabilities and services. Collectively, the three levels of training will provide the strong knowledge basis to make the cultural change necessary to embed data science fully into agency operations. These training tracks address the needs and interests of core data science staff, their peers and colleagues, and senior managers and executives who must make important decisions regarding resources devoted to data science activities and promoting data science activities, personnel, and a culture that values data science.

**Recommendation Five: DIA should establish a Data Science Roundtable.** As noted in recommendation one above, achieving the cultural change to firmly embed data science into military intelligence operations and management will require a coordinated agency-level approach. As described in Chapter Seven, a DSR or similar entity will bring together senior DIA managers with the knowledge and authority to achieve agency-level consensus on data science needs—expertise, capabilities, resources, tradecraft, training, and manpower—and to mobilize resources for meeting those needs. Conceptually, the DSR is a collective decisionmaking body. Establishing a DSR and giving it the authority and responsibility to recommend decisions would bypass contention over authority, responsibility, manpower, and resources that would inevitably be involved in subordinating a data science capability to an existing organization, moving manpower and resources, and reassigning authorities. In the case of the DSR, no authorities or resources move. Elements with a stake in data science decisions would have the opportunity to identify a senior representative to serve on the DSR to represent specific interests. Existing data science activities would be undisturbed, and issues concerning potential new data science activities, especially those requiring cross-agency expertise, could be decided at the DSR level.

The DSR would create an environment that is friendlier to data science and allow the activity to develop and mature within DIA while minimizing contention over division of responsibilities, resources, and authorities. Over time, the DSR could be the leader for organizational change and learning. Similar to the Center of Excellence described above, the DSR could be established for a limited time, with its possible extension based on reaching predetermined measures of effectiveness.

More significantly, the DSR would serve as the basis on which broad cultural changes can be achieved throughout the organization that ultimately will be necessary if data science techniques and methodologies are to be firmly embedded into agency operations. The interviews we conducted underlined the contradictory pressures many intelligence officers feel today; they understand the need to exploit big data for information and intelligence, but at the same time they do not have the tools or training to do so—and many do not trust the output of data analytics. The DSR could lead the way in developing intelligence tradecraft that intelligence professionals will trust. Until this trust is developed, data science will not become a core element of the intelligence cycle.

## The Way Forward

Creating an effective environment for building an enterprise data science capability is not a simple task. Successful application of data science at a low organizational level is an inefficient and likely insufficient way to build a broadly based, enduring capability, and grafting data science onto existing processes is unlikely to grow into a larger analytic methodology and tradecraft. Similarly, trying to build an entirely new data science structure will have significant challenges: The immaturity of the methodology will inhibit its widespread adoption, the need to hire expertise must compete with other resource priorities, and cultural barriers to "new" processes will undermine change. Creating an enterprise-wide data science capability will take time; the obstacles are systemic, and the cultural attitudes and behaviors longstanding.

What is abundantly clear is that strong, persistent, and unified leadership is essential to creating and sustaining an environment in which data science can flourish. Leadership commitment and advocacy for data science must be continuous and visible to analysts, managers, and consumers. Strong leadership will be insufficient unless intelligence producers are recognized and incentivized for the innovative application of data science.

Progress will be slow, and changing years of history, tradition, and culture will be hard. Analysts will initially be especially wary of incorporating data analytics into finished products because they will not immediately trust data whose derivation and authority are not clear. This argues for simultaneously initiating the use of data science projects and identifying experts to develop methods and techniques to qualitatively measure the output of data science analysis. These steps will go a long way toward embedding data science into routine intelligence operations.

Organizations in the private sector face similar challenges as they develop a corps of experts to support their operations. Some organizations have recognized the complexities of this challenge and have made the decision to establish the role of a chief data scientist. A chief data scientist is distinct from a chief data officer: The chief data scientist uses and interprets the organization's data, while the chief data officer owns, regulates, and manages data. DIA may ultimately consider creating this position. A chief data scientist would be directly accountable to the DIA leadership and would clearly demonstrate senior-level support for data science throughout the agency. This step would give strength to leadership support for its commitment to data science by having data science represented at the senior organizational management level. A chief data scientist would be viewed as understanding both the operational output of data science and how to advocate for the resources and organizational placement that data science needs if it is to become a recognized part of the organization's operations and culture.

The reality is that the amount of data will continue to grow. The promise of using big data to monitor international defense and military situations for significant and potentially threatening anomalies and to improve predictive analysis make a compelling argument for adopting this know-how into intelligence analytic tradecraft.

# Interview Protocol

Top of Form

**INTERVIEW ID:**
**INTERVIEW DATE:**
**INTERVIEW TIME:**
**INTERVIEWER NAME:**
**NOTE TAKER NAME:**

## Defining the Roles, Responsibilities and Function of DIA Data Scientists

### Explanation of this project and interview protocol:

Hello, our names are [NAME, POSITION]. We are researchers from RAND, a private non-provider research organization with offices in Santa Monica, Washington DC, and Pittsburgh. RAND has been asked by DIA to conduct a study to determine how best to incorporate data science techniques and data scientists into the DIA workforce. The purpose of this study is to determine what the tasks of data scientists are in the Agency and to recommend how these professionals should be organized to support DIA's collection and analysis efforts. The principal investigators for this project are Brad Knopp and Sina Beaghley, senior analysts at RAND's Washington, D.C. office. [Present copies of business cards.]

During this interview we will ask a series of questions about your use of data science/views on the use of data science in support of your current duties. We specifically want to know about your perceptions of data science and how it is used or not used to support the DIA missions. We will ask for specific vignettes, or stories to help us exemplify the key issues associated with developing a corps of data scientists in the Agency.

The information you provide during this interview will be kept strictly confidential. This interview will last approximately one hour. This interview is anonymous and all of your personally identifiable information will be removed from our data. Your responses will be combined with the responses of all other interviewees to inform our research findings. The data and finished research provided to the U.S. Army will not identify you in any way.

Your participation should cause no risk to you. However, your participation in this interview is strictly voluntary, so if you prefer not to answer a question, or if you want to end this interview for any reason – just let us know. We will not disclose to anyone who did or did not choose to answer individual questions or to participate in the interview. If you do not want to participate, you are welcome to leave now or you can sit here with us for the next hour or so.

*If you have any questions or concerns about the study, we can provide you with the number for the RAND Human Subjects Protection Committee.*

You can contact Brad Knopp at 703 413 1100 x5585 (bknopp@rand.org) or Sina Beaghley at 703 413 1100 x 5203 (beaghley@rand.org), the RAND principal investigators. You can contact the RAND Human Subjects Protection Committee at 310 393 0411.

## 1.0 BACKGROUND INFORMATION

We'd like to talk about data science and its use in intelligence analysis, but before we begin:

Bottom of Form
1.1 Could you briefly describe the output of your organization and the skill level of manpower you use to accomplish that output?

1.2 IF NEEDED: Are your billets intelligence-coded billets (0132) or non-intelligence-coded billets?

## 2.0 WHAT IS DATA SCIENCE?

Thanks very much for that. Let's talk a bit about data science. When we refer to data science, we are talking about the extraction of information from large volumes of structured and/or unstructured data (for example, analyzing large amount of text obtained from social media sites).

2.1 When you think about data science, what tasks or techniques come to mind?

2.2 IF NEEDED: How would you define data science? A data scientist?

2.3 IF NEEDED: What are the primary components of data science?

2.4 In your opinion, how does data science differ from other intelligence techniques?

## 3.0 ANALYSIS

Let's talk about the work you do and the techniques you use.

3.1 How do you think data science should be used in your organization's work?

3.2 How is data science incorporated into the work of your organization?

3.3 What structured techniques do your employees to use in their work? Do those techniques include the product of a data scientist?

3.4 IF NEEDED: How did you come to require use data science products?

3.5. If you have specific data science billets in your organization, did that billet come from outside your organization or did you convert an existing billet?

3.6 Can you describe the process for receiving or requesting the support of a data scientist or receiving data science services?

3.7 What data sets are used for the work? Are there data sets that are not available or are hard to access?

3.8 What was the original justification for incorporating data science or data scientists into your organization?

3.9 Do other organizations in your community use data science techniques?

3.10 Can you describe where you believe data science techniques are most valuable to intelligence professionals?

## 4.0 TRAINING

4.1 What training, if any, have you and your staff received in data science techniques?

4.2 Do you have an official data science certification? Is certification a requirement for data scientists in your organization?

### 5.0 VIGNETTES

5.1 Can you tell me about any specific instances where data science contributed directly to a successful intelligence product? [MORE IS BETTER – MORE DETAIL IS BETTER]

5.2 Conversely, are you aware of instances where data science techniques were not used successfully? Please tell me about them.

### 6.0 DATA SCIENCE AS A CAREER FIELD

6.1 Some IC organizations are creating a career field for data scientists. How do you think this should be done in DIA? Should data science be a separate track? Should data science be another task for all source analysts? Something else?

6.2 Are there intelligence disciplines where data science is more efficiently used than in others?

### 7.0 GENERAL THOUGHTS ON OSINT

7.1 Do you feel data science provides added value to your work? What do you believe are the future possibilities for use to data science in intelligence production?

7.2 How do you think the use of data science is perceived within DIA? Within the Intelligence Community at large?

### 8.0 ENDING

Thank you for your time and for your candid support. Please let us know if there is anything you just said that you would like to have removed from our notes. *[Provide business cards and reiterate rules for informed consent, anonymity, and data storage.]*

# Notional Data Science Position Descriptions

## Notional Position Description: DATA COMMUNICATIONS SPECIALIST

**Position Title:** DATA COMMUNICATIONS SPECIALIST
**Occupational Group:** Intelligence – Analysis
**Mission Category:** Analysis & Production
**Occupational Specialty:** Sci and Eng Analysis
**Pay Plan-Series-Grade:** GG/0000/09
**Work Category:** Professional
**FLSA:** Exempt
**Work Level:** Entry/Developmental
**Job Code Number:** IA0000

### General Summary

The incumbent serves as a Data Communications professional responsible for communicating and presenting summaries of structured and unstructured data in visual, text-based and interactive formats to nonspecialists. This requires strong written, oral, and graphical communication skills. The incumbent develops static visualizations and graphics of complex and big data sets, produces dynamic dashboards for monitoring analytic measures and discovering patterns and anomalies in data, provides analytic narratives and stories to explain the meaning and implications of quantitative analysis, and provides transparency into complex analytic processes by explaining how quantitative and computational methodologies gather and manipulate data in support of analytic, collection, and managerial activities. This position emphasizes a need for an artistic mind to conceptualize, design, and develop reusable graphic/data visualizations that viewers may interact with and be dynamically updated with new information. This position also requires very strong technical knowledge for implementing such data visualizations using the very latest technologies.

## Major Duties

Develops analytic products that are interactive and dynamic to allow users to interact and explore patterns, relations, and links between assessments and underlying data

Explains analytic methods and data collection and processing treatments to make analytic practice transparent and understandable to nonspecialists

Promotes the benefits of data-driven analysis through the production of visualization products

Ability to work well with cross-functional teams and varied communication styles; ability to thrive in a team-oriented environment

Designs and develops analytic dashboards

Tailors the presentation of data analysis to different user communities based on different terminology, priorities and interests

Translates data-driven products into analytic narratives and stories.

Designs and develops finished intelligence products that are interactive and dynamic, changing in response to consumer's interests, needs, and platforms (desktop, tablet, phone, etc.)

Performs other duties as assigned

## Work Category Description

Professional: Positions with duties and responsibilities that primarily involve professional or specialized work that requires the interpretation and application of concepts, theories, and judgment. At a minimum, all groups in this category require either a bachelor's degree or equivalent experience for entry. However, some occupations in this category have positive education requirements (i.e., a requirement for a particular type or level of academic degree). This work category features multiple career progression stages and work levels.

## Work Level Description

Entry/Developmental: In both the Professional and the Technician/Administrative Support Work Categories, work at this level includes learning and applying basic procedures and acquiring competencies through training and/or on-the-job experience. Positions in the Technician/Administrative Support Work Category at this level may involve independent performance of duties. Technician/Administrative Support positions should be placed in this work level when their primary function is the execution of established procedures and standard program practices, and when typical career patterns for the occupation do not extend to the complexity, variety, and scope of the Full Performance Work Level.

## FACTOR A – Knowledge

*Note: Sample relevant academic/professional certifications*
Scientific communication, graphic arts, visual facilitation
Computer science, statistics, informational visualization

## FACTOR B – Guidelines

## FACTOR C – Scope of Authority and Effect of Decisions

## FACTOR D – Work Relationships

## FACTOR E – Supervision Received

## Competencies

Core

Leadership

Occupational

Specialty

# Notional Position Description: DATA ANALYST

**Position Title:** DATA ANALYST
**Occupational Group:** Intelligence – Analysis
**Mission Category:** Analysis & Production
**Occupational Specialty:** Sci and Eng Analysis
**Pay Plan-Series-Grade:** GG/0000/13
**Work Category:** Professional
**FLSA:** Exempt
**Work Level:** Full Performance
**Job Code Number:** IA0000

## General Summary

The incumbent serves as a Data Analyst who is responsible for providing descriptive statistics, probability models, and other quantitative assessments of raw, processed, and generated data. The incumbent employs a combination of traditional statistical and machine learning/artificial intelligence techniques to analyze big and/or complex data sets in support of analytic, collection, and managerial activities. The incumbent works with a variety of databases and types and is familiar with multiple quantitative and computational methodologies. The incumbent creates tailored analyses for a variety of different users in support of collection, analysis, and managerial activities and decisionmaking.

## Major Duties

Performs Exploratory Data Analysis to support hypothesis generation and theory development

Develops and interprets statistical models for hypothesis testing

Work with subject-matter experts to develop metrics or signals to be identified in data and design experiments

Application of supervised and unsupervised machine learning techniques

Work with numerical and non-numerical data, such as text, speech and video

Apply data analytic techniques to specific nuances associated with the collection, handling and analysis of data from different INTs and regions/targets

Address different analytic needs, such as time-series analysis, dimensional reduction, regression, classification, process models, network analysis, etc.

## Work Category Description

Professional: Positions with duties and responsibilities that primarily involve professional or specialized work that requires the interpretation and application of concepts, theories, and judgment. At a minimum, all groups in this category require either a bachelor's degree or equivalent experience for entry. However, some occupations in this category have positive education requirements (i.e., a requirement for a particular type or level of academic degree). This work category features multiple career progression stages and work levels.

## Work Level Description

Work at this level involves independently performing the full range of non-supervisory duties assigned to the employee. Employees at this level have successfully completed required entry-level training or developmental activities either within the employing organization or prior to joining the organization. Employees at this work level have a full understanding of the technical or specialty field, independently handle situations or assignments with minimal day-to-day instruction or supervision, and receive general guidance and direction on new projects or assignments. Within established priorities and deadlines, Full Performance employees exercise independent judgment in selecting and applying appropriate work methods, procedures, techniques, and practices in accomplishing their duties and responsibilities. Actions at this level may have impact beyond the work unit and, as a result, employees at this level typically collaborate internally and externally with their peers.

## FACTOR A – Knowledge

*Note: Sample relevant academic/professional certifications*
Computer science, physics, statistics, econometrics, social network analysis
Applied math, operations research, optimization, informatics
Data wrangling/munging[1]

## FACTOR B – Guidelines

## FACTOR C – Scope of Authority and Effect of Decisions

## FACTOR D – Work Relationships

## FACTOR E – Supervision Received

## Competencies

Core

Leadership

Occupational

Specialty

## Notional Position Description: COMPUTATIONAL SOCIAL SCIENTIST

**Position Title:** COMPUTATIONAL SOCIAL SCIENTIST
**Occupational Group:** Intelligence – Analysis
**Mission Category:** Analysis & Production
**Occupational Specialty:** Sci and Eng Analysis

---

[1]   "Munging" refers to a set of data-related activities, such as data cleaning; combining data sets; and rescaling, normalizing, or otherwise transforming data for use in analysis.

**Pay Plan-Series-Grade:** GG/0000/14
**Work Category:** Professional
**FLSA:** Exempt
**Work Level:** Senior
**Job Code Number:** IA0000

## General Summary

The incumbent serves as a senior Computational Social Scientist and technical adviser, translating social science theory into computational algorithms and developing prototype models for simulating social behavior and processes and analyzing data on social structures and interactions. As such, the incumbent provides methodological support to regional and functional analysts, assists in the development and application of quantitative and computational analytic methods, produces finished analysis based on quantitative and computational methods, and supports decisionmakers through the generation and evaluation of tailored data sets.

### Major Duties

Operationalizes methods and concepts from complex systems/complex adaptive systems

Explains assumptions embedded in statistical analysis methods and their suitability or problems as applied to social systems and individual, organizational, or collective behavior

Employs agent-based modeling, game theory, social network analysis, geospatial analysis, evolutionary computation

Works with subject-matter experts (SMEs) to design constructive models and simulations as advanced structured analytic techniques

Communicates processes and standards for model verification, validation, and accreditation of theoretical and empirical models

Collaborates with analysts and scientists from a variety of disciplinary backgrounds to develop and assess new methodological approaches to addressing the organization's operational, analytics, and organizational needs

Participates in and directs interdisciplinary research teams composed of regional or functional SMEs, physical scientists, engineers, and other professional disciplines

Performs other duties as assigned

### Work Category Description

Professional: Positions with duties and responsibilities that primarily involve professional or specialized work that requires the interpretation and application of concepts, theories, and judgment. At a minimum, all groups in this category require either a bachelor's degree or equivalent experience for entry. However, some occupations in this category have positive education requirements (i.e., a requirement for a particular type or level of academic degree). This work category features multiple career progression stages and work levels.

## Work Level Description

Senior: Work at this level involves a wide range of complex assignments and non-routine situations that require extensive knowledge and experience in the technical or specialty field. Receiving broad objectives and guidelines from the supervisor, senior employees independently handle a wide-range of complex assignments and non-routine situations and exercise independent judgment to identify and take alternative courses of action. Following broad objectives and guidelines, employees act independently to establish priorities and deadlines within expectations established by the supervisor and exercise individual judgment to choose alternative guidelines to complete assignments. Employees may lead and coordinate special projects, teams, tasks, and initiatives and may be required to build and utilize collaborative networks with key contacts within and outside their immediate organization. Actions at this level are likely to have an impact beyond the employee's immediate organization.

## FACTOR A – Knowledge

*Note: Sample relevant academic/professional certifications*
Computational social science, economic, sociology, anthropology, political science, international relations, psychology
Geography, urban studies, cultural studies, ecology, biology, distributed artificial intelligence
Familiarity with one or more social science or life science disciplines
Understands core social science concepts of rationality, organization, authority, identity, risk and uncertainty

## FACTOR B – Guidelines

## FACTOR C – Scope of Authority and Effect of Decisions

## FACTOR D – Work Relationships

## FACTOR E – Supervision Received

## Competencies

Core

Leadership

Occupational

Specialty


## Notional Position Description: DATA ENGINEER

**Position Title:** DATA ENGINEER
**Occupational Group:** Intelligence – Analysis
**Mission Category:** Analysis & Production
**Occupational Specialty:** Sci and Eng Analysis
**Pay Plan-Series-Grade:** GG/0000/15
**Work Category:** Professional
**FLSA:** Exempt
**Work Level:** Expert
**Job Code Number:** IA0000

### General Summary

The incumbent serves as a Senior Data Engineer and technical adviser, working with multiple types of databases, on matters related to capturing and processing live, streaming, and distributed data. As such, the incumbent designs and develops customized data collection, management, and search-and-retrieval systems in order to support the collection, processing, exploitation, analysis and dissemination of big and complex datasets. Also, responsible for supporting the acquisition and development of enterprise level computational resources and customized tools for use by individual or small groups of analysts, collectors, and enabling staff members. The incumbent focuses on the development of software systems, but works closely with the chief information officer's office in order to articulate requirements for specialized hardware and other infrastructure needs.

### Major Duties

Integrates and merges the contents of disparate databases into new datasets that meet the specified needs of users

Processes, generates and translates metadata in order to make data searchable and discoverable for users

Establishes and maintains specialized data collection and processing infrastructure to support specialized mission and user needs

Interacts with the Internet of Things by capturing and merging relevant sources of data for real-time and offline processing

Develops and manages safeguards to protect data holdings according to law, regulation, and policy requirements

Performs other duties as assigned

## Work Category Description

Professional: Positions with duties and responsibilities that primarily involve professional or specialized work that requires the interpretation and application of concepts, theories, and judgment. At a minimum, all groups in this category require either a bachelor's degree or equivalent experience for entry. However, some occupations in this category have positive education requirements (i.e., a requirement for a particular type or level of academic degree). This work category features multiple career progression stages and work levels.

## Work Level Description

Expert: Work at this level involves an extraordinary degree of specialized knowledge or expertise to perform highly complex and ambiguous assignments that normally require integration and synthesis of a number of unrelated disciplines and disparate concepts. Employees at this level set priorities, goals, and deadlines, and make final determinations on how to plan and accomplish their work. Components rely on employees at this level for the accomplishment of critical mission goals and objectives and, as a result, employees may lead the activities of senior and other expert employees, teams, projects, or task forces. Employees at this level create formal networks involving coordination among groups across the intelligence community and other external organizations.

## FACTOR A – Knowledge

*Note: Sample relevant academic/professional certifications*
Computer engineering, databases, big data

Data wrangling/munging, knowledge management/engineering, information technology

## FACTOR B – Guidelines

**FACTOR C – Scope of Authority and Effect of Decisions**

**FACTOR D – Work Relationships**

**FACTOR E – Supervision Received**

**Competencies**

Core

Leadership

Occupational

Specialty

# Methodology

## Structured Data Analysis

The formal analysis of data science education was performed through use of a principal components analysis (PCA) and clustering within networks. By analyzing data from the academic data science programs discussed in Chapter Three, these methods were able to provide a quantitative, visual characterization of data science education and practice.

### Principal Components Analysis

PCA is a statistical technique designed to project high-dimensional, correlated data into lower, uncorrelated dimensions. In practical terms, it is a method for finding structures and relationships in data that may not be apparent when many variables or measurements are addressing the same features. By projecting many correlated variables into a new space, new insights may be gained from the data that would otherwise remain hidden or difficult to detect.

PCA was used to analyze the relationships between data science terms and concepts based on their correspondence with the descriptions of academic course descriptions. This initial assessment proceeded by placing the data into a Burt table or correspondence matrix in which each course was a row in the matrix, and each data science term was a column.[1] In this data structure, each cell contained a zero or one based on whether the course title or description contained the associated data science term; a value of one meant the term was present, and a value of zero meant the term was not. The result was a high-dimensional structure that mapped each course description into a space defined by data science terms and concepts. A sample of the processed Burt table showing the correspondence between academic course descriptions and data science terms is shown in Table C.1.

The resulting data structure provided a high-dimensional representation of the current state of data science education regarding what topics are taught to students

---

[1] For example, see Dell Software, "Correspondence Analysis: Burt Tables," web page, updated May 8, 2015.

**Table C.1**
**Sample Burt Table of Data Science Courses and Terms**

| Course School and Course/ Data Science Term | Agent-Based Modeling | Algorithms | Analysis of Variance | Artificial Intelligence | Autocorrelation | Bayesian Analysis | Big Data | Data Analysis | Data Mining | Database | Decisionmaking | Decision Trees | Distributed Computing | Distributed Database | Empirical | Estimation | Machine Learning | Marlow Chain |
|---|---|---|---|---|---|---|---|---|---|---|---|---|---|---|---|---|---|---|
| Carnegie Mellon University Text Analytics | 0 | 1 | 0 | 0 | 0 | 0 | 0 | 1 | 0 | 0 | 0 | 0 | 0 | 0 | 0 | 0 | 0 | 0 |
| DePaul Introduction to Image Processing | 0 | 1 | 0 | 0 | 0 | 0 | 0 | 0 | 0 | 0 | 0 | 0 | 0 | 0 | 0 | 0 | 0 | 0 |
| DePaul Mining Big Data | 0 | 1 | 0 | 0 | 0 | 0 | 1 | 0 | 0 | 0 | 0 | 0 | 0 | 0 | 0 | 0 | 1 | 0 |
| DePaul Monte Carlo Algorithms | 0 | 1 | 0 | 0 | 0 | 0 | 0 | 0 | 1 | 0 | 0 | 0 | 0 | 0 | 0 | 0 | 1 | 0 |
| George Mason University Complex Adaptive Systems in Public Policy | 1 | 0 | 0 | 0 | 0 | 0 | 0 | 0 | 0 | 0 | 0 | 0 | 0 | 0 | 1 | 0 | 0 | 0 |
| Harvard Introduction to Statistical Modeling | 0 | 0 | 1 | 0 | 0 | 0 | 0 | 0 | 0 | 0 | 0 | 0 | 0 | 0 | 0 | 0 | 0 | 0 |
| Harvard Machine Learning and Data Mining | 0 | 1 | 0 | 0 | 0 | 0 | 1 | 0 | 1 | 0 | 0 | 1 | 0 | 0 | 0 | 0 | 1 | 0 |
| Harvard Monte Carlo Methods for Inference and Data Analysis | 0 | 1 | 0 | 0 | 0 | 1 | 0 | 1 | 0 | 0 | 0 | 0 | 0 | 0 | 0 | 0 | 1 | 1 |
| Indiana Social Media and Language | 0 | 0 | 0 | 0 | 0 | 0 | 0 | 0 | 0 | 0 | 0 | 0 | 0 | 0 | 0 | 0 | 0 | 0 |
| Indiana Statistical Learning and High-Dimensional Data Analysis | 0 | 0 | 0 | 0 | 0 | 0 | 0 | 1 | 0 | 0 | 0 | 0 | 0 | 0 | 0 | 0 | 0 | 0 |
| Indiana Topics in Artificial Intelligence | 0 | 0 | 0 | 1 | 0 | 0 | 0 | 0 | 0 | 0 | 0 | 0 | 0 | 0 | 0 | 0 | 0 | 0 |
| Indiana Topics in Scientific Computing | 0 | 0 | 0 | 0 | 0 | 0 | 0 | 0 | 0 | 0 | 0 | 0 | 0 | 0 | 0 | 0 | 0 | 0 |
| Indiana Web Mining | 0 | 1 | 0 | 0 | 0 | 0 | 0 | 0 | 0 | 0 | 0 | 0 | 0 | 0 | 0 | 0 | 1 | 0 |
| Northwestern Acquisition and Assessment | 0 | 0 | 0 | 0 | 0 | 0 | 0 | 0 | 0 | 0 | 0 | 0 | 0 | 0 | 0 | 0 | 0 | 0 |
| Northwestern Advanced Modeling Techniques | 0 | 1 | 0 | 0 | 0 | 0 | 0 | 0 | 0 | 0 | 0 | 0 | 0 | 0 | 0 | 0 | 1 | 0 |
| Northwestern Analytical Consulting Project Leadership | 0 | 0 | 0 | 0 | 0 | 0 | 0 | 1 | 0 | 0 | 0 | 0 | 0 | 0 | 0 | 0 | 0 | 0 |
| Northwestern Analytics for Big Data | 0 | 1 | 0 | 0 | 0 | 0 | 1 | 1 | 0 | 1 | 0 | 0 | 1 | 0 | 0 | 0 | 0 | 0 |
| Northwestern Analytics for Competitive Advantage | 0 | 0 | 0 | 0 | 0 | 0 | 0 | 1 | 0 | 0 | 0 | 0 | 0 | 0 | 0 | 0 | 0 | 0 |
| Northwestern Big Data Management and Analytics | 0 | 0 | 0 | 0 | 0 | 0 | 1 | 1 | 0 | 1 | 0 | 0 | 0 | 1 | 0 | 0 | 0 | 0 |
| Northwestern Big Data Management/Analytics | 0 | 0 | 0 | 0 | 0 | 0 | 1 | 1 | 0 | 1 | 0 | 0 | 0 | 1 | 0 | 0 | 0 | 0 |
| Northwestern Data Science/Machine Learning | 0 | 1 | 0 | 0 | 0 | 0 | 0 | 0 | 1 | 0 | 0 | 0 | 0 | 0 | 0 | 0 | 1 | 0 |

**Table C.1—Continued**

| Course School and Course/ Data Science Term | Agent-Based Modeling | Algorithms | Analysis of Variance | Artificial Intelligence | Autocorrelation | Bayesian Analysis | Big Data | Data Analysis | Data Mining | Database | Decisionmaking | Decision Trees | Distributed Computing | Distributed Database | Empirical | Estimation | Machine Learning | Marlow Chain |
|---|---|---|---|---|---|---|---|---|---|---|---|---|---|---|---|---|---|---|
| Northwestern Time Series and Forecasting | 0 | 0 | 0 | 0 | 1 | 0 | 0 | 0 | 0 | 0 | 0 | 0 | 0 | 0 | 0 | 0 | 0 | 0 |
| New York University (NYU) Active Portfolio Managment | 0 | 0 | 0 | 0 | 0 | 1 | 0 | 0 | 0 | 0 | 0 | 0 | 0 | 0 | 0 | 0 | 1 | 0 |
| NYU Artifical Intelligence | 0 | 0 | 0 | 1 | 0 | 0 | 0 | 0 | 0 | 0 | 0 | 0 | 0 | 0 | 0 | 0 | 1 | 0 |
| NYU Bayesian Inference and Statistical Decision Theory | 0 | 0 | 0 | 0 | 0 | 1 | 0 | 0 | 0 | 0 | 1 | 0 | 0 | 0 | 1 | 0 | 0 | 0 |
| NYU Big Data | 0 | 0 | 0 | 0 | 0 | 0 | 1 | 0 | 0 | 0 | 0 | 0 | 0 | 0 | 0 | 0 | 0 | 0 |
| NYU Experimental Design | 0 | 0 | 1 | 0 | 0 | 0 | 0 | 0 | 0 | 0 | 0 | 0 | 0 | 0 | 0 | 0 | 0 | 0 |
| NYU Factor Analysis and Structural Equation Modeling | 0 | 0 | 0 | 0 | 0 | 0 | 0 | 0 | 0 | 0 | 0 | 0 | 0 | 0 | 0 | 0 | 0 | 0 |

(based on available course descriptions).[2] This high-dimensional structure was then analyzed based on the properties of its variance through the performance of PCA. In simple terms, the motivation for using the PCA was based on the notion that there are subfields or specializations within data science, and the course descriptions capture these specialties by the way in which they cover and group student exposure to concepts, methods, tools, and applications. By creating composite, abstract variables from the Burt table of associations between data science courses and terms, the underlying structures of data science as a discipline were identified.

Of the 645 courses examined, 598 provided descriptions that contained more than one term from the data science dictionary.[3] Of the more than 4,000 terms in the data science dictionary, 470 appeared in more than one course description.[4] Because the PCA methodology sought to identify an underlying structure within data science based on the association of terms and courses, terms for which there existed no variance were required

---

[2]  A more-complete analysis would examine course syllabi that contain more-detailed information, including outlines of weekly lessons, reading materials, and project criteria that are often absent from shorter descriptions. However, the majority of academic programs do not make detailed syllabi available online.

[3]  Several courses contained no relevant data science terms, as they contained descriptions focused solely on procedural matters devoted to independent research projects, group projects to be performed with local businesses, or references to special topics that were not explicitly identified in the description.

[4]  It is important to consider that the dictionary was developed based on aggregating terminology from several disciplines, not all of which were directly relevant to data science. For example, several terms in the field of computer science pertain to aspects of formal logic or electrical engineering that have limited or no occurrence in data science.

# A Profile of Herbert Simon

The logical and practical fusion of organizational behavior, cognitive science, and artificial intelligence can be seen in the career of Herbert Simon, the pioneering political scientist who developed the theory of bounded rationality. Simon's research focus was on bounded rationality, which described the ways in which innate cognitive processes within the human brain and social organizations limited the ability for individuals and groups to use all available information and consider all possible actions, and therefore could not act in such a fashion as to meet the behavioral criteria of the "rational actor" posited in economics. Instead, people and organizations had to be selective in their use of information and evaluation of plans, due to the high costs of information collection, limited memory, and difficulties associated with evaluating options. Simon argued that individuals and organizations satisficed and employed heuristics in order to guide their search for solutions to problems, terminating their efforts once a satisfactory solution was found; even if it was not an optimal one.

In developing and testing his theory of bounded rationality, Simon turned to the computer as both a model of information processing and an experimental platform for investigating heuristics and simulating cognitive and organizational process. Through the development of these computational research techniques, Simon became a central figure in the development of artificial intelligence and cognitive science. The impact of Simon's research was profound, and the diversity of accolades he received demonstrated the fruitful contributions he made, spanning the social sciences and computer science. For example, he and his collaborator, Allen Newell, received the Turing Award in 1975 from the Association of Computing Machinery, the Nobel Prize in Economic Sciences in 1978, the National Medal of Science in 1986, Institute of Operations Research and Management Science von Neumann Theory Prize, the American Psychological Association Lifetime Achievement Award in 1993, and the American Political Science Association Waldo Award in 1995.a In addition, Simon's work on organizational theory and behavior led him to design the first quantitative programs for business and industrial management at Carnegie Tech's (now Carnegie Mellon University) Graduate School of Industrial Administration.b

[a] Hunter Heyck, "A. M. Turing Award: Herbert ('Herb') Alexander Simon," web page, Association of Computing Machinery, n.d.

[b] Hunter Heyck, *Herbert A. Simon: The Bounds of Reason in Modern America*, Baltimore, Md.: Johns Hopkins University Press, 2005, pp. 145–148.

to be dropped from the dataset. In practical terms, this eliminated terms that were not mentioned in any course description and thus eliminated columns in which all values were zero and had no variation.[5] Likewise, because of the objective of identifying the structure of relations between data science terms and concepts, courses and terms that formed isolated singletons in the network were also eliminated from the dataset. For example, if a term and a course were only associated with one another and did not share any correspondence with any other term or course, they were identified and removed from the data set because their separation from the larger network meant they provided no information about the larger, more complex network of data science associations.

An immediate insight that revealed the breadth and interdisciplinary character of data science education was the fact that 470 terms qualified for inclusion in the final dataset used in the analysis. As noted previously, the initial dictionary consisted of more than 4,000 terms from several different disciplines. That nearly 10 percent of these terms drawn from computer science, statistics, complex systems, and social science were present in a representative sample of courses and programs indicates data science's diversity, multifaceted intellectual origins and research, and professional orientation.

The results of the PCA (see Figure C.1) provided an important image of data science that differed from the traditional characterization offered in academic programs. Specifically, data science practitioners and programs regularly emphasize data science as the combination of computer science, statistics, and visualization. However, the PCA identified a more-complex structure that differed in important respects to the more popular intuitive characterization. The in-depth discussion of the outcome of the PCA can be found in Chapter Four. The PCA assisted in identifying four major categories in data science to help inform potential categories one might find in a team of data scientists.

## Network Clustering

By viewing data science as a network, a distinctive picture emerged that is consistent with the PCA results described earlier. The network structure is depicted in Figure C.2. This figure shows a network based on associations between data science terms according to their colocation in academic course descriptions.

The data depicted in Figure C.2 was analyzed using a clustering algorithm, developed by Clauset, Newman, and Moore, that identified five distinct clusters, depicted in Figure C.3.[6]

These five clusters are visually arranged into blocks, according to their respective size of membership, with each box providing a general label that describes the data sci-

---

[5]   While it was theoretically possible to have terms used by all courses, in which case all column values would be a one, no term of that type was identified, even considering the overrepresentation of the term "course" noted earlier.

[6]   Aaron Clauset, M. E. J. Newman, and Cristopher Moore, "Finding Community Structure in Very Large Networks," *Physical Review E*, Vol. 70, No. 6, December 6, 2004.

**Figure C.1**
**Dimensions of Data Science Identified by the Principal Components Analysis**

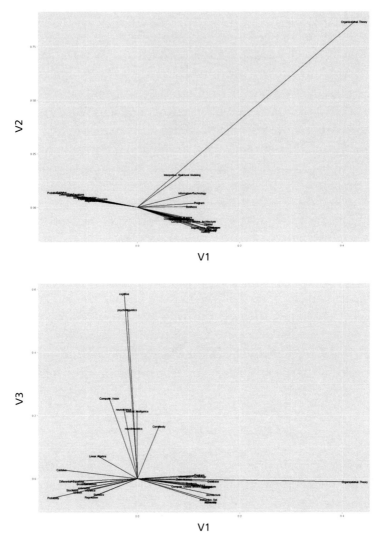

NOTES: The panel on the top shows the associations in the first dimension, V1 on the x-axis, and the second dimension, V2 on the y-axis. Note the two groupings near the origin are on different sides in the first dimension, where terms with a negative value are generally associated with computer science, while terms on the positive side are associated with statistical and mathematical modeling. Organizational theory stands apart as a distinct component in the second dimension. The panel on the bottom depicts the first dimension, V1, on the x-axis and the third dimension, V3, on the y-axis. The general core clusters of computer science and math and statistics remain distinct in both the first and third dimensions. However, the third dimension reveals another grouping of terms associated with neuroscience, cognitive science, and artificial intelligence.

**RAND** *RR1582-C.1*

**Figure C.2**
**Data Science Terms Depicted as a Network Based on Associations Identified Through Correspondence with Academic Course Descriptions**

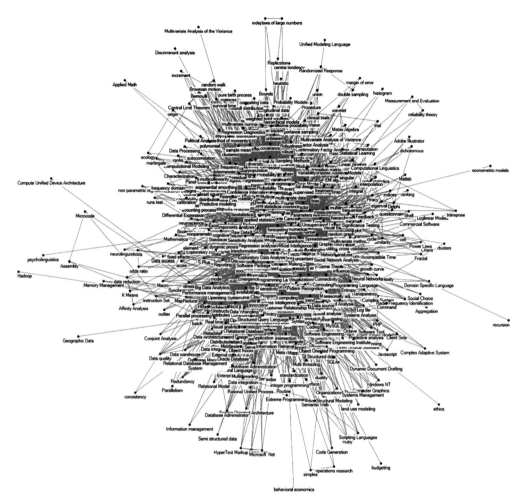

NOTE: Graph generated using NodeXL Pro software, developed by the Social Media Research Foundation.
RAND RR1582-C.2

**Figure C.3**
**Depiction of Data Science Education as a Network of Terms**

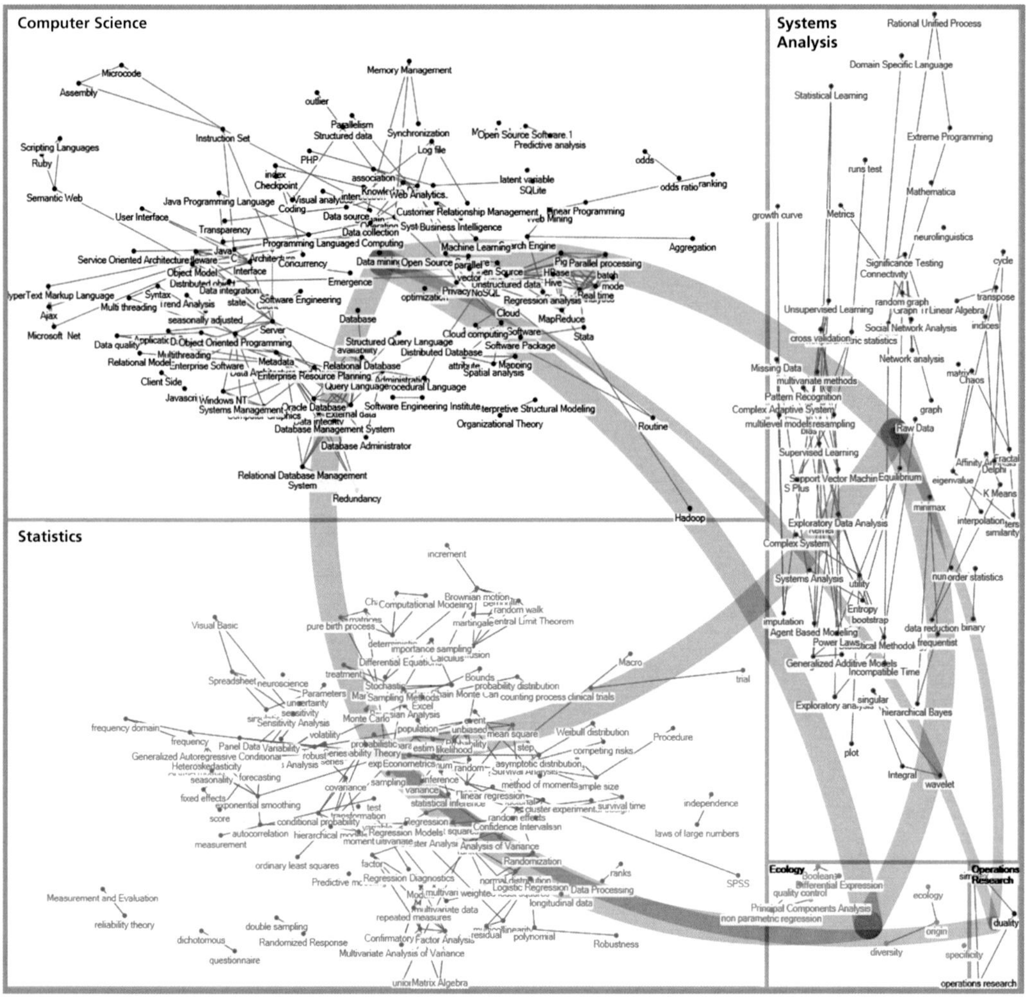

NOTES: While the specific terms are difficult to read, the structure of the graph reveals how data science divides into several distinct structures or communities. These communities may be characterized as computer science, statistics, systems analysis or systems sciences, ecology and biology, and operations research. Graph generated using NodeXL Pro software, developed by the Social Media Research Foundation.

RAND *RR1582-C.3*

ence theme of each grouping. The top-left (largest) group is dominated by computer science terms, while the bottom left is dominated by statistics and mathematics. The top-right box covers conceptual and mathematical tools associated with systems theory and quantitative research methodologies. The two smaller boxes on the bottom cover operations research (bottom far right) and ecology, epidemiology, and computational biology (bottom center right).

As was the case with the dimensional analysis based on the results of the PCA, computational social science emerges as a logical byproduct of the final three, smaller clusters of data science concepts identified in the network of associated terms. Linking the systems analysis and sciences with the study of ecosystems and biological systems underscores the importance and distinctiveness of viewing the worlds of politics, economics, and society broadly as dynamic, adaptive, and evolving. Likewise, the cluster of operations research highlights the importance of merging data science—broadly defined—with decision support and the management of organizational processes and operations.

Finally, the identification of computational social science as a distinctive subfield within data science fulfills early speculation on the future of science broadly defined offered by Warren Weaver in the aftermath of World War II. In 1947, Weaver noted that statistical mechanics had largely solved the problem of large numbers of independent variables that were effectively disorganized or connected randomly, i.e., disorganized complexity, but that organized complexity, cases where variables in systems were structured and interdependent, remained a fundamental challenge to science. Weaver's belief was that the combination of computers, still in their infancy, and interdisciplinary research teams, patterned off of operations research teams that supported military planners during the war, would provide the technical and organizational blueprint for the future of science and overcome the limitations of tools and practices developed to deal with simpler systems.[7] Today, many data science practitioners in academia and business have encouraged their students and staff to become familiar with economics and sociology and view the social sciences as a fertile ground for data science's application.[8] In fact, several data scientists have adopted the term "computational social scientist" when working with large-scale social data sets, such as social media, or developing complex simulation models of individual and collective behavior in financial markets, regional economies, military combat, and more.[9]

---

[7] Warren Weaver, "Science and Complexity," *American Scientist*, Vol. 36, No. 4, October 1948, pp. 536–544. Also see John von Neumann and Oskar Morgenstern, *Theory of Games and Economic Behavior*, Princeton, N.J.: Princeton University Press, 2007; and Simon A. Levin, "Complex Adaptive Systems: Exploring the Known, the Unknown, and the Unknowable," *Bulleting (New Series) of the American Mathematical Society*, Vol. 40, No. 1, October 2002, pp. 3–19.

[8] Gutierrez, 2014, pp. 9–10, p. 65, and p. 103.

[9] For references to computational social science as an academic discipline, see David Lazer, Alex Pentland, Lada Adamic, Sinan Aral, Albert-Laszlo Barabasi, Devon Brewer, Nicholas Christakis, Noshir Contractor, James Fowler, Myron Gutmann, Tony Jebara, Gary King, Michael Macy, Deb Roy, and Marshall Van Alystne, "Computational Social Science," *Science*, Vol. 323, No. 5915, 2009, pp. 721–723; Stanford University, "Institute for Research in the Social Sciences Center for Computational Social Science," web page, n.d.; Computational Social Science Society of the Americas, home page, n.d; and George Mason University, "Krasnow Institute for Advanced Study—Computational Social Science," web page, n.d.

# References

AnalyticBridge, "Data Science Dictionary," blog post, November 17, 2012. As of May 4, 2015: http://www.analyticbridge.com/profiles/blogs/2004291:BlogPost:223153

Arrow, Kenneth J., Robert Forsythe, Michael Gorham, Robert Hahn, Robin Hanson, John O. Ledyard, Saul Levmore, Robert Litan, Paul Milgrom, Forrest D. Nelson, George R. Neumann, Marco Ottaviani, Thomas C. Schelling, Robert J. Shiller, Vernon L. Smith, Erik Snowberg, Cass R. Sunstein, Paul C. Tetlock, Philip E. Tetlock, Hal R. Varian, Justin Wolfers, and Eric Zitzewitz, "The Promise of Prediction Markets," *Science*, Vol. 320, No. 5878, May 16, 2008, pp. 877–878.

Atwood, Chandler P., "Activity-Based Intelligence: Revolutionizing Military Intelligence Analysis," *Joint Force Quarterly*, Vol. 77, April 1, 2015, pp. 24–33.

Berkeley School of Information, "What Is Data Science?" web page, n.d. As of January 20, 2016: https://datascience.berkeley.edu/about/what-is-data-science/

Borne, Kirk, "Big Data A to ZZ—A Glossary of My Favorite Data Science Things," Converge Blog, March 21, 2014. As of May 4, 2015: https://www.mapr.com/blog/big-data-zz-%E2%80%93-glossary-my-favorite-data-science-things

Brown, Bobbi, "A Best Way to Manage a CMS Hospital Readmission Reduction Program," web page, HealthCatalyst, n.d. As of May 4, 2015: https://www.healthcatalyst.com/healthcare-data-warehouse-hospital-readmissions-reduction

Buckingham, Marcus, and Ashley Goodall, "Reinventing Performance Management," *Harvard Business Review*, April 2015, pp. 40–50.

Centers for Medicare and Medicaid Services, "Readmissions Reduction Program," web page, updated August 4, 2014. As of May 4, 2015: http://www.cms.gov/Medicare/Medicare-Fee-for-Service-Payment/AcuteInpatientPPS/Readmissions-Reduction-Program.html

Central Intelligence Agency, "Careers and Internships: Data Scientist," web page, March 10, 2016. As of June 27, 2016: https://www.cia.gov/careers/opportunities/science-technology/data-scientist.html

Clauset, Aaron, M. E. J. Newman, and Cristopher Moore, "Finding Community Structure in Very Large Networks," *Physical Review E*, Vol. 70, No. 6, December 6, 2004.

ComplexityBlog.com, "Glossary of Terms," blog post, n.d. As of May 4, 2015: http://complexityblog.com/resources/glossary.html

Computational Social Science Society of the Americas, home page, n.d. As of July 6, 2016: https://computationalsocialscience.org/

Daft, Richard, *Management*, Boston: Cengage Learning, 2009.

Data Analytics & R, "What Is. . . ," blog post, n.d. As of July 6, 2016:
https://advanceddataanalytics.net/what-is/

DataFloq, "An Extensive Glossary of Big Data Terminology," web page, n.d. As of May 4, 2015:
https://datafloq.com/abc-big-data-glossary/

DataInformed, "Analytics and Big Data Glossary," web page, updated
September 24, 2014. As of May 4, 2015:
http://data-informed.com/glossary-of-big-data-terms/

Davis, Jessica, "Intel Chief Data Scientist Shares Secrets to Successful Projects," *Information Week*,
November 16, 2015. As of January 22, 2016:
http://www.informationweek.com/big-data/big-data-analytics/
intel-chief-data-scientist-shares-secrets-to-successful-projects/d/d-id/1323145

Dell Software, "Correspondence Analysis: Burt Tables," web page, updated
May 8, 2015. As of January 22, 2016:
http://documents.software.dell.com/Statistics/Textbook/Correspondence-Analysis#burt

Deshpande, Bala, "Connecting Dots: Preventive Maintenance, Big Data, Internet of Things," blog
post, Simafore.com, October 16, 2013. As of May 4, 2015:
http://www.simafore.com/blog/bid/119414/
Connecting-dots-preventive-maintenance-big-data-internet-of-things

Dhar, Vasant, "Data Science and Prediction," *Communications of the ACM*, Vol. 56, No. 12,
December 2013, pp. 64–73.

Dutcher, Jenna, "Sentiment Analysis Symposium: Uncovering Human Motivations," blog post,
*datascience@berkeley*, March 7, 2014. As of May 4, 2015:
http://datascience.berkeley.edu/sentiment-analysis-symposium-uncovering-human-motivations

Elster, Jon, *Explaining Social Behavior: More Nuts and Bolts for the Social Sciences*, New York:
Cambridge University Press, 2007.

EMC Digital Universe with Research and Analysis by IDC, "The Digital Universe of Opportunities:
Rich Data and the Increasing Value of the Internet of Things," web page, April 2014. As of
January 20, 2016:
http://www.emc.com/leadership/digital-universe/2014iview/executive-summary.htm

George Mason University, "Krasnow Institute for Advanced Study—Computational Social Science,"
web page, n.d. As of July 6, 2016:
http://www.css.gmu.edu/

Gild, "SocialCode + Gild: Using Gild to Discover Talent Who Are 'Hidden Gems' for a Startup with
a Unique Culture," web page, n.d. As of May 4, 2015:
https://www.gild.com/customers/socialcode-case-study/

Giuliani, Matteo, Andrea Castelletti, Francesca Pianosi, Emanuele Mason, and and Patrick M. Reed,
"Curses, Tradeoffs, and Scalable Management: Advancing Evolutionary Multi-Objective Direct
Policy Search to Improve Water Reservoir Operations," *ASCE Journal of Water Resources Planning
and Management*, Vol. 142, No. 2, 2014.

Goertz, Gary, *Social Science Concepts: A User's Guide*, Princeton, N.J.: Princeton University Press, 2012.

Goldstein, Jeffrey, "Resource Guide and Glossary for Nonlinear/Complex Systems Terms," web page,
PlexusInstitute.org, n.d. As of May 4, 2015:
http://c.ymcdn.com/sites/www.plexusinstitute.org/resource/resmgr/files/
goldstein_-_resource_guide_a.pdf

Grimm, Volker, and Steven F. Railsback, *Individual-Based Modeling and Ecology*, Princeton, N.J.: Princeton University Press, 2005.

Gutierrez, Sebastian, *Data Scientists at Work*, New York: Apress, 2014.

Hadka, David M., Jonathan Herman, Patrick Reed, and Klaus Keller, "OpenMORDM: An Open Source Framework for Many-Objective Robust Decision Making," *Environmental Modeling & Software*, Vol. 74, 2014, pp. 114-129.

Heyck, Hunter, "A. M. Turing Award: Herbert ('Herb') Alexander Simon," web page, Association of Computing Machinery, n.d. As of January 22, 2016:
http://amturing.acm.org/award_winners/simon_1031467.cfm

———, *Herbert A. Simon: The Bounds of Reason in Modern America*, Baltimore, Md.: Johns Hopkins University Press, 2005.

———, *Age of System: Understanding the Development of Modern Social Science*, Baltimore, Md.: Johns Hopkins University Press, 2015.

Hochster, Michael, "What Is Data Science? And What Is It Not?" *Quora*, January 16, 2014. As of January 20, 2016:
https://www.quora.com/What-is-data-science

IBM Software, *Reducing Hospital Readmissions for Congestive Heart Failure*, 2012.

Inter-University Consortium for Political and Social Research, "Summer Program in Quantitative Methods of Social Research," web page, n.d. As of January 22, 2016:
https://www.icpsr.umich.edu/icpsrweb/sumprog/index.jsp

Karckhardt, David, and Jeffery R. Hanson, "Informal Networks: The Company Behind the Chart," *Harvard Business Review*, Vol. 71, 1993, pp. 104–111.

Kasprzyk, Joseph R., Patrick Reed, and David M. Hadka, "Battling Arrow's Paradox to Discover Robust Water Management Alternatives," *ASCE Journal of Water Resources Planning and Management*, Vol. 142, No. 2, 2014.

KDnuggets, "Data Mining and Predictive Analytics Glossary," blog post, n.d. As of May 4, 2015:
http://www.kdnuggets.com/2015/06/data-mining-predictive-analytics-glossary.html

Kelly-Detwiler, Peter, "Machine to Machine Connections—the Internet of Things—and Energy," *Forbes*, August 6, 2013.

Kobielus, James, "Data Scientists: Grow and Sustain a Center of Excellence," blog post, *IBM Big Data & Analytics Hub*, May 21, 2012. As of January 22, 2016:
http://www.ibmbigdatahub.com/blog/data-scientists-grow-and-sustain-center-excellence

Kotadia, Harish, "Key Big Data Terms You Should Know," blog post, April 9, 2013. As of May 4, 2015:
http://hkotadia.com/archives/5427

Lazer, David, Alex Pentland, Lada Adamic, Sinan Aral, Albert-Laszlo Barabasi, Devon Brewer, Nicholas Christakis, Noshir Contractor, James Fowler, Myron Gutmann, Tony Jebara, Gary King, Michael Macy, Deb Roy, and Marshall Van Alystne, "Computational Social Science," *Science*, Vol. 323, No. 5915, 2009, pp. 721–723.

Leek, Jeff, "The Key Word in 'Data Science' Is Not Data, It Is Science," blog post, *Simply Statistics*, December 12, 2013. As of January 20, 2016:
http://simplystatistics.org/2013/12/12/the-key-word-in-data-science-is-not-data-it-is-science/

Lempert, Robert J., Steven W. Popper, and Steven C. Bankes, *Shaping the Next One-Hundred Years: New Approaches to Long-Term Policy Analysis*, Santa Monica, Calif.: RAND Corporation, MR-1626-RPC, 2004. As of May 4, 2015:
http://www.rand.org/pubs/monograph_reports/MR1626.html

Levin, Simon A., "Complex Adaptive Systems: Exploring the Known, the Unknown, and the Unknowable," *Bulleting (New Series) of the American Mathematical Society*, Vol. 40, No. 1, October 2002, pp. 3–19.

Los Alamos National Laboratory, "GENetic Imagery Exploitation," web page, 2011. As of June 22, 2016:
http://www.genie.lanl.gov/

Murray, Adrienne, "What Is the Value of Healthcare Dashboards?" web page, HealthCatalyst, August 8, 2013. As of May 4, 2015:
https://www.healthcatalyst.com/value-of-healthcare-dashboards

New York University, "What Is Data Science?" web page, n.d. As of January 20, 2016:
http://datascience.nyu.edu/what-is-data-science/

Okasha, Samir, *Philosophy of Science: A Very Short Introduction*, New York: Oxford University Press, 2002.

O'Leary, Daniel, "Prediction Market as a Forecasting Tool," *Advances in Business and Management Forecasting*, Vol. 8, 2011, pp. 169–184.

Palmer, Shelly, "Data Science 101: Definitions You Need to Know," blog post, September 7, 2014. As of May 4, 2015:
http://www.shellypalmer.com/2014/09/data-science-101/

Patil, D. J., "The Importance of Taking Chances and Giving Back," in Carl Shan, Henry Wang, William Chen, and Max Song, eds., *The Data Science Handbook*, Middletown, Del.: The Data Science Bookshelf, 2015, pp. 16–27.

Pentland, Alex, "The New Science of Building Great Teams," *Harvard Business Review*, April 2012, pp. 60–70.

"Predicting Hospital Readmissions: Laura Hamilton Interview (Additive Analytics CEO)," *Data Science Weekly*, n.d. As of May 4, 2015:
http://www.datascienceweekly.org/data-scientist-interviews/predicting-hospital-readmissions-laura-hamilton-interview-additive-analytics-ceo

Press, Gil, "A Very Short History of Data Science," *Forbes*, May 28, 2013. As of January 20, 2016:
http://www.forbes.com/sites/gilpress/2013/05/28/a-very-short-history-of-data-science/#2715e4857a0b57661cfe69fd

"Q&A: Cathy Johnston," *Geospatial Intelligence Forum*, Vol. 13, No. 2/3, 2015. As of January 20, 2016:
http://www.kmimediagroup.com/gif/articles/424-articles-gif/q-a-cathy-johnston/6850-q-a-cathy-johnston

Railsback, Steven F., and Volker Grimm, *Agent-Based and Individual-Based Modeling: A Practical Introduction*, Princeton, N.J.: Princeton University Press, 2012.

RAND Corporation, "Robust Decision Making," web page, n.d. As of May 4, 2015:
http://www.rand.org/topics/robust-decision-making.html

Reuschemeyer, Detrich, *Usable Theory: Analytic Tools for Social and Political Research*, Princeton, N.J.: Princeton University Press, 2009.

Santa Fe Institute, "Complex Systems Summer School," web page, n.d. As of January 22, 2016:
http://santafe.edu/education/schools/complex-systems-summer-schools/

Schrieber, Jared M., *The Application of Prediction Markets to Business*, thesis, Cambridge, Mass.: Massachusetts Institute of Technology, 2004.

Seibel, Fred, and Chuck Thomas, "Manifest Destiny: Adaptive Cargo Routing at Southwest Airlines," Uncluttered.com, 2000. As of May 4, 2015:
http://www.unclutteredcom.com/portfolio_materials/manifestdestiny.pdf

Singh, R., Patrick Reed, and K. Keller, "Many-Objective Robust Decision Making for Managing an Ecosystem with a Deeply Uncertain Threshold Response," *Ecology and Society*, Vol. 20, No. 3, 2014.

Statistics.com, "Glossary of Statistical Terms," web page, n.d. As of May 4, 2015:
http://www.statistics.com/glossary/

Southern Methodist University, "Data Science @ SMU," web page, n.d. As of January 22, 2016:
https://cdn3.datascience.smu.edu/

Stanford University, "Institute for Research in the Social Sciences Center for Computational Social Science," web page, n.d. As of July 6, 2016:
https://iriss.stanford.edu/css

Suroweicki, James, *The Wisdom of Crowds*, New York: Anchor, 2005.

Thomas, Chuck R., Jr., and Fred Seibel, "Adaptive Cargo Routing at Southwest Airlines," Ernst & Young Center for Business Innovation, 1999. As of
January 22, 2016:
https://www.academia.edu/11158857/Adaptive_Cargo_Routing_at_Southwest_Airlines

Various authors, Quantitative Applications in the Social Sciences, series, Thousand Oaks, Calif.: SAGE Publications, 1987–2014.

von Neumann, John, and Oskar Morgenstern, *Theory of Games and Economic Behavior*, Princeton, N.J.: Princeton University Press, 2007.

Weaver, Warren, "Science and Complexity," *American Scientist*, Vol. 36, No. 4, October 1948, pp. 536–544.

Wilensky, Uri, and William Rand, *An Introduction to Agent-Based Modeling: Modeling Natural, Social, and Engineered Complex Systems with NetLogo*, Cambridge, Mass.: MIT Press, 2015.